HAVE YOU TRIE[D]
TO LOSE W[EIGHT?]

This book was written because most diets have not been very successful. Many are simply too weak to get results. Others use an artificial "formula approach" that no one could live with permanently.

Protein is the subject of many myths. The bottom line on protein is this: High-protein diets are dangerous. Many formula diets emphasize high-protein foods and contain very little carbohydrates. This type of diet is not a formula for success! It can cause a rapid, and usually temporary water loss. But generally the weight comes back on very quickly.

There is a much better way.

—Neal Barnard, M.D.

FOODS
THAT CAUSE
YOU TO
LSE
WEIGHT

The Negative Calorie Effect

NEAL BARNARD, M.D.

wm

WILLIAM MORROW
An Imprint of HarperCollins*Publishers*

This book contains advice and information relating to health care. It is not intended to replace medical advice and should be used to supplement rather than replace regular care by your doctor. It is recommended that you seek your physician's advice before embarking on any medical program or treatment. All efforts have been made to assure the accuracy of the information contained in this book as of the date of publication. The publisher and the author disclaim liability for any medical outcomes that may occur as a result of applying the methods suggested in this book.

HarperCollins books may be purchased for educational, business, or sales promotional use. For information please e-mail the Special Markets Department at SPsales@harpercollins.com.

FIRST WILLIAM MORROW PAPERBACK PRINTING: JULY 2016

FIRST HARPER MASS MARKET PRINTING: MAY 2011

FIRST AVON BOOKS MASS MARKET PRINTING: MARCH 2002

FIRST WHOLECARE MASS MARKET PRINTING: APRIL 1999

Library of Congress Cataloging-in-Publication Data has been applied for.

ISBN 978-0-06-257036-9

16 17 18 19 20   DIX/RRD   10 9 8 7 6 5 4 3 2

CONTENTS

Contents

MESSAGE FROM
NEAL BARNARD, M.D.

Thank you for purchasing this revolutionary book. You are going to love it. Even more, you are going to love the way you feel in the new body that is waiting for you. As you read, you may find yourself thinking that this looks much too easy. We designed it to be as simple and user-friendly as possible. In fact, it looks so easy that you may be tempted not to try it. People who have struggled with weight problems year after year sometimes feel that if a diet is not punishing, it could not possibly work.

Try this program for just twenty-one days. If your results are like most other people, you will be able to watch yourself getting thinner. You can actually chart your progress week by week. This program is not a diet. It is a delicious way to shift your menu from fat-adding foods to foods that actually increase your metabolism so that the cells of your body burn calories faster. These foods are naturally modest in calories. They cannot go straight to fat and actually increase your body's calorie-burning rate. You have many of these foods already in your cupboard.

The trick is to know which ones they are and how to maximize their action.

This effect can be very powerful. A woman from the Midwest called my office not too long ago. She had seen an interview I had done on television and wanted to let me know how this simple plan had worked for her. She had struggled with a weight problem all of her life. "I've been on every diet there is," she said, "and the weight kept coming back, and then some. For me, it just kept getting worse and worse." At first, she was a bit timid about trying this program. After all, she had been on so many diets that had promised results, only to let her down again. In addition, because my program suggests greatly increasing foods that step up the body's calorie burning, and preferring these foods to others (which do not help the body burn calories), she was not sure if she really wanted to try different foods.

However, she decided to try the program for three weeks, knowing that if she did not like the results, she could stop. She found it hard to believe that she did not need to count calories or limit her portion size. As you will see, this approach encourages you to eat normal portions at every meal, which is a terrific advantage over old-fashioned diets. She also really loved a couple of my soups and entrees, and brought them to work to share with friends. Sometimes, after eating one of the meals from the recipe section, she felt a slight warm sensation, as if her body really was burning calories faster. In fact, that is exactly what was happening.

After a week, she stepped on the scale. Her weight had

dropped three pounds. Could be a fluke, she thought. After all this did not feel like a diet at all. But she kept it up. The next week she continued to lose weight. And the following week, she dropped even more. However, a funny thing was happening. Her tastes began to change. She started to prefer the fat-burning foods and no longer desired many of the fattening foods she had been accustomed to.

After six months, she could not believe her success. She had lost more than 100 pounds. Now, I do not necessarily recommend losing weight that fast. It is best to let it come off gradually, but that gives you an idea of the power of this approach.

When you try this simple program, you not only will take into your hands the most effective permanent weight management tool ever devised, but you will also have an understanding of powerful weight management techniques that most other people have never even heard of.

Try this easy, step-by-step method.

1. Read Part One of this book. This short section, called "Basic Concepts," will give you all the information you need to start.
2. Now you will have a good grasp of how to proceed. Take a look at Part Two, "Let's Get Started." As you do so, imagine all the foods that help boost your calorie-burning (you now know what they are), and choose which ones you would like to focus on for the next three weeks. Many people prefer spaghetti with a rich tomato sauce, rice pilaf, or bean burritos. You do not have to

stick to just the recipes you find here. Go ahead and use any others that fit the guidelines you have read.

3. As you begin, familiarize yourself with Part Three, "Moving into High Gear." Like the preceding two sections, it is very short, and easy to read and understand. Then flip through the recipes in Part Four, and see which ones you would like best.

I do not ask you to stick to this program forever. Just try it for three weeks. During that time, give it your best effort. Take your picture before you start and jot down your starting weight.

Each week, record your new weight in a diary. If you are like most people, you will feel better, you will have more energy, and your weight will have started to come down on its own without attempting to control the size of your meals. If you like what is happening—and I think you will—stick with it. Let the pounds come off on their own, you are going to love the way you feel and look. You will probably find that you will want to stay with your new way of eating. I have been doing exactly this for many years, and I am in better shape and have more energy now than ever before. Many other people are enjoying exactly the same wonderful experience.

As you progress, track your success and add a picture of yourself to your diary every two months. When you have finished your diary, send it to me. I am interested to know how it has worked for you.

In case you are wondering how all this came about, let

me tell you a little about myself. I graduated from medical school in 1980 at George Washington University School of Medicine in Washington, D.C., where I also did my residency. Then I moved to New York to work at St. Vincent's Hospital, which is a large medical complex in downtown Manhattan.

Afterwards, I decided to move back to Washington, D.C., and form an organization called the Physicians Committee for Responsible Medicine. I now divide my time between my duties as President of PCRM, and lecturing and writing to spread information about health to as many people as possible. You may have seen me or other PCRM doctors on *60 Minutes, Oprah, Good Morning America*, or other programs. It is particularly wonderful that television networks have taken such an interest in this topic. You can write to me through Avon Books, 1350 Avenue of the Americas, New York, New York, 10019 or PCRM in Washington, D.C.

A FEW BEGINNING TIPS

- Let the fat melt away at its own speed. Do not push for overly quick weight loss.
- If you want to exercise, I strongly suggest keeping it simple. Don't overdo it. A daily half-hour walk or an hour walk three times per week is plenty

for starters. If you are over forty, out of shape, or have any medical condition, see your doctor first. Some people cannot exercise because of arthritis or other conditions. The food changes you are making will help you burn calories faster *whether you exercise or not.* So while exercise is great and can add even more to your calorie burning, bring in the foods that help you burn fat first, and increase your physical activity only gradually and when you are ready. Again, calorie-burning foods help you whether you exercise or not.

- As powerful as this program is, it is general information and it does not take the place of individualized medical care. You should have a physician of your own, especially if you are ill or on medication. Let your doctor know that you are beginning this program, in case he/she needs to adjust medication doses. For example, people with high blood pressure or diabetes often improve with these dietary changes, so their doctors will need to reduce or eliminate their medications.

This is exactly what happened to a man from St. Louis, Missouri. He had had a weight problem for many years. He also had a high cholesterol level, and had been on insulin for adult-onset diabetes. He decided to try the guidelines you will read about. The results were decisive. He lost seventy pounds. However, that was not all. His cholesterol dropped more than eighty points. His blood sugar fell back to normal, so his doctor stopped his insulin. He simply did not need it anymore. All of these improvements stayed with him, and he is now a picture of health and energy.

I hope you love your new knowledge and your new way of selecting foods. You can use this program as much or as little as you like, but the more your meals fit into the simple guidelines you will soon learn, the more power you have to be in the body you want.

Let me wish you the very best of success!

Sincerely,

Neal D. Barnard, M.D.

INTRODUCTION

This book offers a new approach to weight control. If your goal is a slimmer body and more energy than you have had in years, this program is more powerful than any diet you have ever tried. It is not a diet. It is a comprehensive program that brings about better weight control than old-fashioned diets ever could.

In this program, we will not compromise. We will use all the factors that are known to promote permanent healthful weight control. You will boost your metabolic rate through food selection, shift your menu from calorie-dense fatty foods to foods with a much better nutrient makeup, and bring in other ways to burn calories more effectively.

This book was written because most diets have not been very successful. Many are simply too weak to get results. Others use an artificial "formula approach" that no one could live with permanently. They may cause a phenomenal weight loss for a few weeks, followed by a huge weight gain back to and beyond the starting point. These frustrating results are caused by the poor design of the diets.

There is a much better way. For years, researchers have used many different kinds of diets to improve health: diets to help people lose weight, lower their cholesterol levels, or deal with various health problems. From these studies, it is clear that there are certain factors that are critical to success in long-term weight control.

For example,

- Certain foods add easily to fat stores while others do not.
- Certain foods will alter your metabolism so that calories are burned more effectively.
- Weight loss should be gradual, not sudden.
- And, most importantly, a menu that is very low in fat, high in fiber and carbohydrate, and modest in protein, is extraordinarily powerful for weight control.

We will examine these and many other factors in detail.

Old diets focused mainly on quantity. In the new approach used in this program, the type of food that makes up the menu is considerably more important than the quantity of food. In other words, *you can eat until you are full and still lose weight.* No more skimpy portions. If you find that difficult to believe, consider the research. Many clinical studies have examined diets that are based on very small portions and other diets that change the type of food but not the quantity. It turns out that *the type of food you eat is much more important for permanent weight control than the portion size.* In fact, skimpy portion diets can actually impair your ability to burn off calories. Research also shows that thin people actually eat more food, not less, compared to most overweight people. In this program, we will shift the diet toward foods that boost the metabolism and that do not easily add to body fat.

There are also specific factors that make the process of change easier. A diet that brings about long-term success is much more rewarding than one that yields only modest or short-term results. So this program is designed to be as effective as possible.

We will also have opportunities to try new kinds of foods. This is not only fun; it is the critical step in building a new body. Tasting food is when theories about weight control turn into practice. You will learn about specific new foods to try, including menus, recipes, and shopping lists. In addition, we will look at ways to bring in family and friends. They are our allies in life's changes. We will see how to help them help us.

When you put this program to work for you, you can have the slimmest body nature ever intended for you.

HOW TO USE THIS PROGRAM

This program is divided into four parts:

Basic Concepts

First, find a comfortable chair and read Part I. In this concise section, you will learn the basic concepts that are essential to a permanent slimming-down program.

Let's Get Started

Part II gets you started with a step-by-step program, including new foods to try, recipes, and a simple program for physical activity. You can easily chart your progress.

Moving into High Gear

Part III moves the program into high gear, with more advanced information to solidify your gains.

Menus and Recipes

Part IV gives you many more healthful recipes to make eating a joy.

Get all you can out of this program. No corners were cut in bringing this information to you; do not cut any corners in putting it into action.

You can succeed. So be patient, and don't rush. It took time for you to gain weight, and it will take time for it to come off again. Overly rapid weight loss can lead to a rebound of weight gain, so let the pounds come off gradually. I think you will be happy with the new you.

A NOTE TO THE READER

My goal is to provide you with information on the power of foods for health. However, neither this book nor any other can take the place of individualized medical care or advice. If you have any medical condition, are overweight, or are on medication, please talk with your doctor about how dietary changes, exercise, and other medical treatments can affect your health.

The science of nutrition grows gradually as time goes on, so I encourage you to consult other sources of information, including the references listed in this volume.

With any dietary change, it is important to insure complete nutrition. Be sure to include a source of vitamin B_{12} in your routine, which could include any common multivitamin, fortified soymilk or cereals, or a vitamin B_{12} supplement of five micrograms or more per day.

I wish you the very best of health.

FOODS THAT CAUSE YOU TO LOSE WEIGHT

PART I

Basic Concepts

BASIC CONCEPTS

We need to think differently about the approach to losing weight. Forget old-fashioned diets. There are very good reasons why they do not work well. Your body was not designed recently. The human body took shape millions of years ago, long before diets were invented. At that time, the lack of food meant only one thing, starvation, and if the body could not cope with the lack of food, the result was life-threatening. So we have built-in mechanisms to preserve ourselves in the face of low food intake. These defenses are automatically put to work. When you go on a low-calorie diet, you know that you are doing so to lose weight. But your body does not know that. As far as your body is concerned you are starving, and it will trigger a number of biological mechanisms to try and stop you.

To see how to avoid this problem, let's first look at how your body burns calories. The speed at which your body burns calories is called the *metabolic rate*. Some people have a "fast metabolism" and burn lots of calories in a short time. They are likely to stay slim. Other people have a slower metabolic rate and have a harder time staying slim.

Your metabolism is like the rate at which an automobile uses up gas. An idling car uses up some fuel. When the car is moving it uses more, and when it accelerates up a hill it will use a lot more gas.

Our bodies work the same way. We burn some calories even when we are relaxing or asleep because it takes energy to maintain our normal body temperature and to keep our lungs, heart, brain, and other organs working. When we engage in activities, the more strenuous they are, the more calories we burn.

DIETING SLOWS
YOUR METABOLISM

The point to remember is that your metabolic rate can be changed. In a period of starvation or dieting, the body slows down the metabolism. The body does not understand the concept of dieting. Remember, as far as your body is concerned, a diet is starvation, and it does not know how long the starvation period will last. So it clings to its fat like a motorist running out of gas preserves fuel. Remember the last time you were driving along the highway and suddenly noticed that the gas gauge was below empty? You tried to remember how far below "E" your gauge will go. You went easy on the accelerator, driving very smoothly, and turned off the engine at stop lights to conserve gas until you got to a station.

Your body does the same sort of thing when food is in short supply. It turns down the metabolic flame to save as much of the fat on your body as possible until the starvation period is over, because fat is the body's fuel reserve. This is very frustrating to dieters. They often find that, even though they are eating very little, their bodies do not easily shed the pounds. Even worse, the slowed metabolism can continue beyond the dieting period, sometimes for weeks, according to studies at the University of Pennsylvania and elsewhere.[1] For that reason, fat is easily and rapidly accumulated again after the dieting period. This causes the familiar yo-yo phenomenon, in which dieters lose some weight, then rebound to a higher weight than they started with.

Here is the first step to keeping your metabolic rate up: Make sure that your diet contains at least 10 calories per pound of your ideal body weight. This means that if you

---

Minimum Calories per Day
Ideal Weight × 10 Calories = Calories

120 × 10 calories = 1200 calories
150 × 10 calories = 1500 calories
180 × 10 calories = 1800 calories

---

are aiming for a weight of 150 pounds, your daily menu should contain at least 1500 calories. Weight loss will be gradual, but you will not slow your metabolism and, so, you will be able to retain your progress.

AVOIDING BINGES

There is another problem with skimpy eating. Not only does the body lower its metabolic flame to conserve energy, but it also gets ready to take maximal advantage of any food source it finds. When food becomes available, there is a tremendous tendency to binge, in what is known as the *restrained-eater* phenomenon. You know the pattern. You have been dieting for several days, and suddenly someone brings home a carton of ice cream. A little bit won't hurt, you decide, and before you know it you are scraping the bottom of the carton and digging around the cracks for every last bit. You then scold yourself for your "lack of will-power." The truth is that the problem was not willpower at all, but the innate biological programming of the human body. The diet turned on the "anti-starvation" plan that is built into every human being. Your body assumed that any food in front of you might be the only calorie source you might have for a while, so it demanded a binge.

> • The point to remember is binges come from diets.

It is not a question of weak will or gluttony. The human body has a built-in tendency to binge after periods of starvation.

For a similar reason, it is best not to skip meals. Skipping breakfast and lunch leads to overeating later in the day. So, eat regular meals and avoid very-low-calorie diets.

Bulimia—binge eating often followed by purging—almost always begins with a diet. And as the binging begins, shame and secrecy often follow. If this has happened to you, remember that binging is not a moral failing. It is a natural biological consequence of dieting.

Dieting is now a nearly universal pastime in America, and bulimia is an ever-growing epidemic. Unfortunately, children are raised on a menu that is almost certain to make many of them gain weight. The cultural trend in western countries in the past several decades has emphasized meat, dairy products, fried chicken, french fries, and other high-fat foods. Combined with an increasingly sedentary lifestyle, the predictable result is that many people will become overweight. They mistakenly believe that the problem is the *quantity* of food they are eating, rather than the *type* of food. Rather than abandon the offending foods, they simply eat less. A restrictive diet begins. The natural result is lowered metabolic rates, cravings, and binging. Most binges would probably never occur if dieting were replaced with better food choices that would promote a slow, steady drop in weight, rather than an overly rapid weight loss.

Skipped meals and skimpy portions are not effective for permanent weight control and are not a part of this program.

AN OPTIMAL
WEIGHT-LOSS MENU

Now that you know what not to do, let's build a program that takes pounds off and keeps them off. The basis of this program is a way of eating that promotes weight control naturally, without counting calories and without skimpy portions.

Let's first look at *carbohydrates*. The starchy white inside of a potato is mostly complex carbohydrate, which is simply a chemist's term for molecules made up of many natural sugars linked together. When you eat a potato, that carbohydrate is gradually broken apart into simple sugars, which are absorbed and used by the body. Rice, wheat, oats, and other grains are rich in complex carbohydrates, as are beans and nearly all vegetables.

In the past, many people believed that starchy foods were fattening. They would avoid carbohydrate-rich potatoes, rice, bread, and pasta. People would take a baked potato, which has only 95 calories, and top it with a pat or two of butter, and sometimes add sour cream, grated cheese, or bacon bits. As they gained weight, they blamed the potato. But we now know that what was fattening was not the potato, but the greasy toppings that were added to it.

- Carbohydrate-rich foods are actually low in calories.
- A gram of carbohydrate (about $\frac{1}{30}$ ounce) = 4 calories.

That is why a slice of bread has only 79 calories and an ear of corn only 120. A chicken breast, which contains no carbohydrate, has fully 86 calories. In contrast, starchy foods are low-calorie foods.

Compare that to fatty foods. A gram of fat has 9 calories, more than twice the calorie content of carbohydrate. It is only when carbohydrate-rich foods are covered with fatty toppings that lots of calories are added.

Look at how this works with actual foods. You are planning a candlelight dinner for two. A nice spaghetti dinner with some fresh vegetables, perhaps a glass of wine. A one-cup serving of spaghetti topped with ½ cup of tomato sauce has about 200 calories. But if we decided to add ground beef to the sauce, look what happens: the spaghetti dinner suddenly has 365 calories. The fat in ground beef holds a lot of calories.

Let's take another example: A half-cup serving of mashed potatoes has 70 calories. A tablespoon of butter on top adds fully 108 calories. In the process a low-calorie food becomes a high-calorie food. In other words, fatty toppings are high in calories, but the carbohydrate-rich potatoes, spaghetti, bread, etc., are not.

Food and Fatty Toppings

Potato (1 med.) = 95 calories
Potato (1 med.) + Butter (1 Tbs) = 203 calories
Potato (1 med.) + Butter (1 Tbs) + Cheese
(1 oz) = 317 calories

There may be an advantage to whole unprocessed grains, such as rice, cereal, or corn, as opposed to grain that has been ground up into flour (e.g., bread or pasta). Some evidence shows that we tend to extract more calories from the ground-up varieties, perhaps because the process of "digestion" has been begun for us.

There are other important virtues of carbohydrates. They cannot add *directly* to your fat stores. We do not have any "carbohydrate storage areas" on our bellies or thighs. If the body is to store the energy of carbohydrates in fat, it has to chemically convert the carbohydrate molecules into fat. This process consumes a fair number of calories. As a result,

- Calories from carbohydrates are not as likely to increase body fat as are the same number of calories from fats.

In addition, *carbohydrates boost your metabolism*. Plant-based meals tend to increase the metabolic rate slightly. Here is how it works. Carbohydrate breaks down in the body to various sugars. Sugars cause insulin to be released which, in turn, leads to the production of two natural hormones, norepinephrine and thyroid hormone (T3). T3 and norepinephrine both increase the metabolic rate. The result is more effective calorie burning.[2–5]

So starchy foods are naturally low in calories, they cannot be automatically added to body fat, and they help boost your metabolic rate so that calories are burned off a bit faster.

- Carbohydrate-rich vegetables, beans, and grains are the best friends of anyone trying to shed some pounds.

Now here is a critical point.

- Complex carbohydrates are found only in plants.

Grains, such as bread, spaghetti, and rice are loaded with carbohydrates. Beans and vegetables are also high in

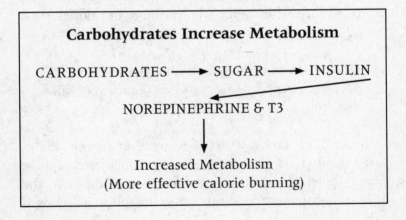

Carbohydrates Increase Metabolism

CARBOHYDRATES ⟶ SUGAR ⟶ INSULIN

NOREPINEPHRINE & T3

↓

Increased Metabolism
(More effective calorie burning)

carbohydrates. But there are virtually no complex carbo-hydrates in chicken, fish, beef, pork, eggs, or dairy prod-ucts. The more animal products you eat, the more you are pushing carbohydrate-rich vegetable foods off your plate. That is one reason why the most effective weight-control programs use vegetarian menus.

There is an added bonus to foods from plants: fiber. Grains, beans, and vegetables contain fiber, which adds texture and makes them filling and satisfying. Fiber is what people used to call roughage, the part of plants that resists digestion in the small intestine. The value of fiber was not appreciated until relatively recently, and so it was often removed by refining methods. The result was white bread instead of whole-grain breads, white rice instead of brown rice, and baked goods that were more densely packed with calories and less satisfying than they would have been had the fiber been left in. Fiber adds a hearty texture to foods but has virtually no calories.

Like complex carbohydrates, fiber is found only in plants. Grains, such as wheat, oats, rye, corn, rice, and the breads, cereals, and other foods that are made from them are loaded with fiber. Vegetables of all kinds and legumes, such as beans, peas, and lentils are also rich in fiber.

Animal products contain no fiber at all. To the extent that animal products are added to the diet, the fiber con-tent is reduced. Americans now consume only 10–20 grams of fiber per day, on average, which is about half of what we should have. The reason, of course, is the pen-chant for animal products and refined plant foods, which unfortunately displace the fiber-rich foods. *But do not feel that you must calculate your fiber intake.* When you center

Charlene: Beating a Weight Problem

Charlene wanted to lose 30 pounds. In fact, she had wanted to lose these same 30 pounds for several years. She had tried several different diets, including some with formula drinks, and had also tried diet pills. None of these were effective over the long run, although all had seemed to help temporarily. When I met her, she was avoiding all carbohydrates. She skipped breakfast, had yogurt and turkey slices for lunch, and usually ate frozen dietetic meals for dinner. Her weight had been essentially the same for months.

I suggested that, instead of avoiding starchy foods, she make them the center of her diet. Breakfast was to be hot cereal and fruit. At work, she could make lunch from dried soups. For dinner, she was to make a pot of rice, as much as she could eat, or, if she preferred, she could have potatoes or other starchy foods instead. She was also to include vegetables and beans or lentils at dinner. Because this "diet" included a rather large quantity of food, she worried that she might actually gain weight on it. But some rather simple calculations showed that the calorie content of this menu was actually very modest. She lost weight very gradually, but about ten months later, the 30 extra pounds were gone.

your diet on high-carbohydrate foods, such as whole grains, beans, and vegetables, the fiber content of your diet will increase naturally. As you will see in Part II, the result will be meals that are satisfying and filling. When we discuss the value of carbohydrate-rich foods and fiber, you can simplify this by thinking in terms of foods from plants versus animal products. A plant-based diet is rich in carbohydrate and fiber.

Animal products are devoid of them. The result is that plant-based diets promote slimness, while animal products promote overweight.

THE NEGATIVE CALORIE EFFECT

Many people still believe that the number of calories in any given food tells you just how fattening that food is likely to be. For example, a cup of rice has about 220 calories. Three slices of bologna also have 220 calories. So some people assume that these two foods have exactly the same effect on the waistline.

They don't. The very same number of calories coming from bologna and from rice have very different effects. The bologna tends to be fattening, as a general rule, while the rice does not.

Rice does provide calories to run the body's functions.

And theoretically it is possible for unused calories from rice to be stored as fat. But it turns out that rice is much less fattening than the same number of calories from bologna, other meats, or other fatty foods. Rice—like other carbohydrate-rich foods—has a way of naturally *reducing* the calories that are available for fat storage.

You might think of this as a "negative calorie effect." One of the most exciting concepts in the science of weight control in many years is the fact that certain foods can actually assist in the *loss* of fat.

By now, it will come as no surprise to you that carbohydrate-rich foods are power foods for weight control. But let's see what the "negative calorie effect" really means. Then, we'll look at twenty foods that encourage this effect and that you can eat freely. In reality there are far more than twenty and by the time you are done with this book, I hope you will have gone far beyond the old-fashioned notion of counting calories and limiting portion size. The key is not how much you eat, but, instead, the types of foods you eat.

When you think of carbohydrate, think, for example, of rice. A rice grain is a seed, designed by nature to start a new rice plant. The starchy white interior of a rice grain consists mainly of complex carbohydrates that nourish the seed as it sprouts and grows. The same is true of beans, potatoes, apples, and many other plants. The starchy carbohydrate interior provides nourishment for the tiny growing plant.

For millions of years, humans and other primates have plucked fruits from trees and roots from the ground and have taken advantage of carbohydrate's capacity to nourish us. What is remarkable is that these foods provide energy with relatively little tendency to cause overweight. In many Asian countries, for example, where rice is still the center of the diet and huge amounts of rice are consumed, people tend to remain slim.

While carbohydrates provide calories for the body, they also have ways of counteracting the storage of some of these calories as fat, and also encourage the burning of stored calories:

First, as we saw earlier, a substantial number of the calories in carbohydrates are used up as carbohydrates are turned to fat. Let me give you some numbers: For every 100 calories of carbohydrate that your body tries to store as fat, 23 are lost in the process of breaking down carbohydrate molecules and building fat molecules from them. That means that, of the 220 calories in a cup of rice, about 50 calories are used up just in the chemical processing. Leaving grains whole, like rice, cereals, or corn, rather than grinding them into flour to make bread or pasta, also causes them to release fewer calories.

But that is just the beginning. In addition, because carbohydrate increases the body's metabolism, more calories are burned off as the metabolism increases. The metabolism-boosting effect causes more of the calories in *all the foods you eat* to be burned. When that happens, they cannot be turned into fat.

It is similar to the effect of turning up a car's idle. More gas is used up, there is less in the tank, less to spill on the ground, and less to use in the future, because it has been burned.

Another part of the "negative calorie effect" of carbohydrates is that they are the part of the diet that tells the body when it has had enough food. Your body does not just pay attention to how much you have eaten. It actually has a way to monitor how much carbohydrate is coming in. When it has had enough, it reduces the feeling of hunger. Carbohydrates are the cue the body needs. So, if there is a lot of carbohydrate on your plate, you will tend to eat to feel satisfied and to turn down the drive to fill your plate. The natural sugar in fruits, called fructose, also has an appetite-reducing effect.

What this means is that if you have included generous amounts of rice, potatoes, beans, fruits, and other carbohydrate-rich foods on your meals, the calories in pork chops, salad oil, and other fattening foods are less likely to find their way onto your fork.

How do you get these "negative calorie effects"? You will not get them from steak or fried chicken, because there is virtually no complex carbohydrate in fish, chicken, beef, milk, eggs, or any other animal product. Complex carbohydrates are found only in plants. Grains, vegetables, and beans are loaded with them. That is why vegetarian foods are such powerful foods for permanent weight control.

If you like, you can forget technical terms like carbohydrate. As long as your diet is made from grains, beans, vegetables, and fruits, rather than animal products, it will be naturally rich in carbohydrate.

20 FOODS YOU CAN EAT IN VIRTUALLY UNLIMITED PORTIONS

Listed below are twenty foods that you should feel free to eat in very generous portions. Unless you are really stuffing yourself, you can eat as much of these as you want. In fact, there are many more than twenty, as you have learned. One caveat: Enjoy these with no butter, margarine, or oily toppings—fats are fattening!

Corn	Celery
Rice	Peas
Potatoes	Cauliflower
Lettuce (all varieties)	Pineapple
Broccoli	Cabbage
Carrots	Oranges
Black beans	Apples
Kidney beans	Grapefruit
Spinach	Bananas
Lentils	Oatmeal

CUTTING OUT FATS AND OILS

Now for the most important part of the food prescription.

- Cut out the fats and oils.
- Fats and oils are packed with calories.
- Fat in foods is fat on you.

These are the most calorie-dense part of the foods we eat. As we noted previously, every single gram of fat or oil holds 9 calories. This is true for all fats and oils: beef fat, chicken fat, fish oil, vegetable oil, and any other kind of fat or oil.

There are various kinds of fat. The main categories discussed by dietitians are *saturated fat*, which is common in animal products and is solid at room temperature, and *unsaturated fat*, which is common in vegetable oils and is liquid at room temperature. Different kinds of fat have different effects on your cholesterol level. But

- For weight control, we need to be concerned about all forms of fat.
- All fats and oils have the same calorie content: 9 calories in every gram.

About 35 percent of the calories most Americans get every day come from fat. For a typical 2000-calorie menu, that is 700 calories each day just from fats and oils in our foods. By cutting out most of the fats in our diet, we can cut out hundreds of calories. To put it another way, if all the foods we eat are very low in fat, we can eat far more food than we could on a high-fat diet, without more calories.

We should cut our fat intake from 35 percent of the calories we eat down to about 15 percent. Eating 15 percent of our calories from fat is a substantial reduction. It is a powerful weight-reducing step and yields other tremendous benefits as well. We must go on a "search and destroy" mis-

Food	Percentage of calories from fat
Potato	less than 1%
Peas	3%
Black beans	4%
Macaroni noodles	4%
Vegetarian baked beans	4%
Rice	less than 5%
Cauliflower	6%
Spinach	7%
Broccoli	8%
Wheat bread	15%
Whole milk	49%
2% milk	35%
Extra lean ground beef	54%
Ground beef	60%

sion for fat. *Be on the lookout for fat in the two forms in which it comes: animal fat and vegetable oil.*

Animal fat was designed by nature to act as a calorie-storage area for animals. When you eat animal fat, you are eating all those stored calories. Animal fat is not only on the outside of a cut of meat. It is marbled through the lean part, too, almost like a sponge holding water. So, if you are eating meat you are eating someone else's fat and someone else's concentrated stored calories. It will put fat on you. Let's take some examples:

Imagine that we are making tacos. Let's compare two recipes for taco filling, one made with ground beef and the other with beans. Beef is high in fat; three ounces of ground beef hold about 225 calories. Beans are very low in fat, and three ounces hold only about 80 calories. So we can cut out nearly two-thirds of the calorie content by switching from the beef recipe to the bean recipe. A big part of the difference is the very high fat content of the ground beef and the very low fat content of beans. About 60 percent of the calories in ground beef come from fat. This is a huge load of calories that do nothing good for the body and do a lot of harm, from promoting heart disease to increasing cancer risk, and, of course, fattening you up.

Take a look at the fat content of various foods in the chart on the previous page. Remember, the fat contents listed here are percentages of calories, not percentages by weight. This is a critical difference. Whole milk, for example, is 3.3 percent fat by weight, because it is loaded with water. But 49 percent of its calories come from fat. Milk that is 2 percent fat by weight is actually about 35 percent fat as a percentage of calories. It is actually not a low-fat product at all.

"Extra lean" ground beef is really not so lean either: it derives 54 percent of its calories from fat. It is an abysmal food for people concerned about their waistlines.

By calorie content, here's a listing of common meat cuts.

Common Meat Cuts

Cut	Percentage of calories from fat
Chuck roast	51%
Rib eye steak	63%
Short loin porterhouse	64%
Hot dogs	82%
Bologna	83%
Most beans, grains, vegetables	less than 10%

Even the beef industry in its "lean" advertisments of the "skinniest six" beef specimens could not find any cuts of meat that are anywhere near the fat content of beans, grains, or vegetables.

All of these have many times the fat content of typical vegetables, beans, grains, and fruits.

The problem with meats, including poultry and fish, is that they are muscles, and muscles are made up of protein and fat. They contain no fiber at all and virtually no carbohydrate.

"Skinniest Six" Meat Cuts

Cut	Percentage of calories from fat
Tenderloin	41%
Top loin	40%
Sirloin	38%
Round tip	36%
Eye of round	32%
Top round	29%
Most beans, grains, vegetables	less than 10%

Advertisers sometimes claim that chicken and fish are low-fat foods. Are they? Let's look at the worst and the best of the poultry line.

The chicken also contributes about 85 mg. of cholesterol. In addition, chicken pushes carbohydrates and fiber off

Poultry	Percentage of calories from fat
Chicken frank	68%
Roasted chicken	51%
White meat w/out skin	23%
Most beans, grains, vegetables	less than 10%

your plate. No matter how chicken is prepared, it cannot get its calorie level down to that of the truly healthful foods, because chicken, like all meats, is permeated by fat and contains no complex carbohydrates or fiber. *Fat always has more calories than carbohydrate.*

Some people eat fish in the hope that fish oil will reduce their cholesterol levels. Actually, fish oils reduce triglycerides but do not reduce cholesterol levels. And it should be remembered that fish oils are as fattening as any other oils or fats. Like all fats and oils, they contain 9 calories per gram.

Different types of fish differ greatly in their fat content.

Fish	Percentage of calories from fat
Chinook salmon	52%
Atlantic salmon	40%
Swordfish	30%
Halibut	19%
Snapper	12%
Sole	9%
Haddock	8%

Many of these are as bad as other animal products. Others are in the same ballpark as vegetables as far as their fat content goes, but this does not make them recommended foods. Remember fish contains no complex carbohydrates and no fiber, and tends to displace these foods from the meal. All fish products also contain cholesterol and far too much protein (see pages 38–42), in addition to contamination problems. So, fish is still not a health food, although certain types of fish are much lower in fat than are beef and poultry.

In summary, meats, poultry, and fish have two main problems for those concerned about their weight.

First, like all muscles, they have inherent fat, adding concentrated calories.

Second, because muscle tissues are mainly just protein and fat, they reduce the carbohydrate and fiber content of the diet. They displace the fiber and carbohydrates that are essential to a satisfying and metabolism-boosting menu.

- The first prescription for cutting the fat is the V-word: vegetarian foods are power foods for weight control.
- The second issue is vegetable oil.

Vegetable oils have received a good reputation because they contain no cholesterol and are low in saturated fats. But their calorie content is the same as any other kind of fat. That should be emphasized. *All fats and all oils, regardless of type (lard, pork fat, chicken fat, olive oil, fish oil, etc.), are packed with calories: 9 calories per gram.* They are all the enemies of those in search of a slimmer waistline.

Let's take an example with vegetable oils. As you know, a potato is a low-fat food that is also modest in calories. Only about 1 percent of the calories in a potato come from fat. When the potato is baked or prepared as mashed potatoes, no extra oil is added. But if the potato is cut into french fries and dropped in cooking oil, its fat content soars up to 40 percent or more. As a result, its calorie content doubles or even triples.

Compare the fat content of a doughnut (50%), which is fried in oil, to a bagel (8%), which is not. The doughnut has more than six times the fat of the bagel.

Fat in foods adds easily to your fat stores.

Little or no conversion is needed in the body before the fat we eat passes through the digestive tract and the bloodstream to the fat tissues of the body. But the energy in carbohydrates cannot be stored easily as fat. The body has to do a considerable amount of work before those calories can be stored, and many calories are lost in the process.

1 Large Raw Potato		2 Regular Fries
70 calories	+ Frying =	440 calories
1 gram fat		23 grams fat

Fats Are Calorie-Dense

Carbohydrate	4 calories per gram
Protein	4 calories per gram
Fat	9 calories per gram

FAT AND OTHER HEALTH CONSIDERATIONS

There are other serious problems with fats, too. Fat in foods contributes substantially to the risk of several forms of cancer (breast, colon, prostate, and others), heart disease, diabetes, gallstones, and numerous other problems as well. Although animal fats are the worst, vegetable oils also increase health problems.

A low-fat menu is a recipe for a slim, healthy body. It can take some getting used to because, unfortunately, people crave high-fat foods. Grease is like an addicting substance. We all have a tendency to return to fried chicken, greasy burgers, potato chips, and fried onion rings, so be on the lookout. It is easier to cut them out entirely than to continually tease oneself with occasional greasy foods.

GETTING FREE FROM
THE FAT IN FOODS

As you know from the above discussion, shifting away from fat and toward high-fiber foods means getting away from animal products. Cutting out animal products can greatly reduce your fat consumption. Avoiding added vegetable oils and fried foods is another very powerful step. Here are some suggestions on how to do it.

Salad Dressings. Salad dressings can be packed with fat. A salad made of one cup of romaine lettuce with half a tomato holds only 20 calories. But look what adding a tablespoon of dressing will do.

Salad dressing (1 Tbsp.)	*Fat content*
Catalina french dressing	5.5 grams of fat (65 calories)
vinegar and oil (50/50 mix)	8.0 grams of fat (72 calories)
Good Seasons Zesty Italian	9.2 grams of fat (85 calories)

So the salad with dressing has four to five times the calories of a salad without dressing.

Low-fat or no-fat dressings cut down substantially on the fat content. Or you might prefer a sprinkle of lemon or lime juice as a dressing for salad or vegetables. A tablespoon of lemon or lime juice has no fat and only 4 calories. You may also find that you enjoy the taste of fresh spinach, chickpeas, tomatoes, or other salad ingredients with no dressing at all.

Baked Goods. In recent years, nutritionists have made distinctions between saturated (animal fat, tropical, and hydrogenated oils) and unsaturated (most vegetable) oils, because the former contribute to heart problems. But if our goal is to slim down, the issue is much simpler: all kinds of fats and oils are problems. They are all packed with calories. Some baked goods, such as bagels, pretzels, and many breads, are usually quite low in fat. On the other hand, croissants, cakes, pies, and cookies tend to be very high in fat. Commercially packaged goods list their ingredients on the label. The ingredients are listed in decreasing order of their quantities, so if oil is one of the first ingredients, there is probably more of it than if it is one of the last listed ingredients. In addition, most labels provide enough information to allow you to calculate the actual fat content, using the simple formula on page 33.

We do need some fat in the diet. But we need only a fraction of what most of us typically get. A small amount of fat is inherent in grains, legumes, and vegetables. This is all the body needs. Children can (and perhaps should) have a bit more fat in their diet. Breast milk is naturally higher in fat for the needs of growing infants. The natural process of weaning eliminates this nutrient when it is no longer appropriate.

FOR MEAT-LOVERS ONLY

To reduce the fat content of the diet, low-fat vegetarian foods are ideal. Vegetarian foods are obviously free of the animal fat that permeates meats, poultry, and fish. Steering clear of fried foods and added oils is the other half of the equation. Spaghetti with tomato sauce, bean burritos, vegetable curries, baked potatoes, and salads are a few examples of foods that can be very low in fat, yet delicious.

As you'll discover in this book, the foods that power up your body's ability to burn calories come from grains, vegetables, beans, and fruits, which is why we have included plenty of recipes that allow you to enjoy these foods as often as possible. My goal is to help you burn off weight as quickly as you can, without ever having a single hunger pang.

"But, wait a minute," you might be saying. "What about the occasional steak or chicken breast? Are they out of bounds with this program?"

Well, it is true that our taste for these and other fatty foods has forced us to cut an extra notch in our belts (not to mention making our doctors worry about our cholesterol levels). However, as a doctor, I recognize that, while many people are ready to eat 100-percent healthful foods three times a day, 365 days a year, you may not be. Some people prefer a more gradual approach.

If this means you, let me encourage you to try this.

- Have as many of the foods as possible from the list on page 19, along with whatever else you may be eating. Have them every day, in as generous portions as possible. A halfway approach is not nearly as good as a complete commitment, but it is a beginning to a slimmer and healthier you.
- Try the recipes. They really are delicious, and when you find the ones you like, add them to your routine. The more of them you have on a regular basis, the more powerful the effect on your waistline.
- Even better, stick to the foods in this program very strictly, but *only do so for three weeks*. This puts the program to a good test and will soon show you why it is as powerful (and popular) as it is. If you like its slimming effect, you can stick with it.

HOW TO CHECK THE
FAT CONTENT OF FOODS

How much fat is in foods we eat?

What is important is the percentage of calories that come from fat. (The percentage of fat by weight is not important, because it can be easily thrown off by the water content of products.)

On the information panel on the package, notice the number of calories from fat in one serving. Then divide by the number of calories per serving. Then multiply by 100.

$$\frac{\text{Calories from fat}}{\text{calories in serving}} \times 100 = \begin{array}{l}\text{\% calories} \\ \text{from fat}\end{array}$$

Let's try an example. Here is an information label from a supposedly low-calorie pizza.

Nutritional Facts

Serving size	1.0 oz	Calories	65
Servings per pkg	8	Calories from fat	23

This serving is incredibly small, so all the numbers will be artificially low. But even so, the key to look for is the percentage of calories that come from fat. If we divide the number of calories from fat (23) by the number of calories per serving (65), we come out with .35. Then multiplying by 100 gives us 35 percent. This means that 35 percent of the calories in this product are from fat. This is better than fried chicken or a hot dog, but still higher than we want. Not such a healthful entree after all.

$$\frac{23}{65} = .35 \times 100 = 35\% \text{ of calories from fat}$$

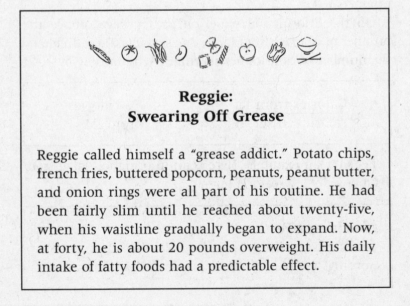

Reggie:
Swearing Off Grease

Reggie called himself a "grease addict." Potato chips, french fries, buttered popcorn, peanuts, peanut butter, and onion rings were all part of his routine. He had been fairly slim until he reached about twenty-five, when his waistline gradually began to expand. Now, at forty, he is about 20 pounds overweight. His daily intake of fatty foods had a predictable effect.

He did not plan to give up any foods forever. But as an experiment, he decided that for three weeks he would eat only low-fat vegetarian meals. From a set of recipes, he made bean entrees with lots of vegetables on the side. There was no limit on amounts, but he was very strict to omit all oils, margarine, salad dressings, and all meats and dairy products. After three weeks, he had lost about three pounds. That was not a big drop, but he found that he had lost all desire for greasy foods, and now associated them with his weight problems. So he decided to stick with his new way of eating for three more weeks. He lost five more pounds. A month later, he had lost another 5 pounds. He now weighs the same as he did in college.

His friend Morris adopted the same program. He weighed 275 pounds when he started. Without limiting calories, he lost 80 pounds. His girlfriend used the same method to drop from 150 to 120 pounds.

CHECK YOUR KNOWLEDGE

Let's review. For each pair below, see if you can pick which is lower in fat. You'll find the answers below. Do not skip this part. It is easy, but important.

Which is lower in fat?

1. Fried chicken vs. broiled top round beef
2. Leanest beef vs. leanest chicken
3. Leanest chicken vs. vegetarian baked beans
4. Leanest beef vs. rice
5. Leanest chicken vs. potato
6. Spaghetti with tomato sauce vs. a Lean Cuisine spaghetti with meatballs dinner
7. Spaghetti with meat sauce vs. spaghetti with tomato sauce
8. Fast-food meat taco vs. fast-food bean burrito
9. Cheddar cheese vs. bread
10. Peanut butter vs. rice
11. Ice cream vs. jelly beans
12. Baked potato vs. french fries
13. Doughnut vs. bagel

Answers

(The numbers given are percentages of calories from fat.)

1. Broiled top round beef (38%) is lower in fat than fried chicken (50% fat), but both are high-fat foods.
2. The leanest chicken is about 20% fat, and lower than the leanest beef (29% fat), although both are high in fat compared to grains, beans, vegetables, and fruits.
3. Vegetarian baked beans (4% fat) are much lower in fat than even the very leanest chicken (20%).
4. Rice (0.8%) is much lower in fat than the leanest beef (29% fat).

5. A potato (1%) is much lower in fat than the leanest chicken (20% fat).

6. Spaghetti with tomato sauce (6%) is much lower in fat than a Lean Cuisine spaghetti with meatballs dinner (23% fat).

7. Spaghetti with tomato sauce (6%) is lower in fat than spaghetti with meat sauce (35%).

8. A fast-food bean burrito (31%) is lower in fat than a fast-food meat taco (50% fat). A home-made burrito can be much lower in fat.

9. Bread (16%) is much lower in fat than cheddar cheese (74% fat). Most cheeses are extremely high in fat.

10. Rice (0.8%) is much lower in fat than peanut butter (78% fat).

11. Jelly beans (0.8%) are much lower in fat than ice cream (48% fat), although both hold a very large amount of sugar.

12. A baked potato (1%) is much lower in fat than french fries (47% fat).

13. A bagel (8%) is much lower in fat than a dough-nut (50% fat).

Carbohydrate-rich foods are vital for long-term weight control. Take the carbohydrate test below.

Which has more carbohydrate?

1. A fish fillet vs. broccoli
2. Bread vs. beef
3. Milk vs. potato
4. Cheese vs. rice

Answers

(The numbers given are percentages of calories from carbohydrate.)

1. Broccoli is 78% carbohydrate. A fish fillet has no carbohydrate at all.
2. Bread is 75% carbohydrate. Beef has no carbohydrate at all.
3. A potato is 93% carbohydrate. Milk is 30% carbohydrate, in the form of simple sugar.
4. Rice has much more carbohydrate (89%) than cheese (1%).

WHAT ABOUT PROTEIN?

Protein is the subject of many myths. The bottom line on protein is this: High-protein diets are dangerous. Many formula diets emphasize high-protein foods and contain very little carbohydrate. This type of diet is not a formula for success. It can cause a rapid, and usually temporary, water loss. But usually the weight comes back on very quickly.

In addition, there are serious dangers to high-protein diets: osteoporosis and kidney disease. The bone-thinning disease of osteoporosis is an epidemic in the U.S., and protein has apparently been a big part of the cause. High-protein diets cause calcium to be lost in the urine. This has been shown repeatedly in scientific studies. When subjects consume foods that are overly high in protein, especially animal protein, they excrete calcium. For example, if volunteers were to eat meals with a substantial meat content and later have their urine tested, calcium would often be found in the urine. Since Americans tend to eat meat daily, it is likely that they are routinely excreting calcium. The calcium does not come from the meat. It comes from their bones.

There are several theories that scientists have used to explain this. The amino acids which make up protein and are released when protein is digested make the blood slightly acidic. In the process of buffering this acid, calcium is pulled from the bones. Ultimately it is discarded in the urine. In addition, meat protein is very high in what are called "sulfur-containing amino acids." These are suspected of being particularly likely to leach calcium from the bones.

While many of us grew up being taught to make sure we got enough protein, the fact is we have gotten too much. Our bodies only need a fraction of what we generally get. When we eat two or three times the amount of protein the body can use, much of it is broken down and excreted. In the process, it not only interferes with the calcium balance of the body, it can also overwork the kidneys. The excess of amino acids acts as a diuretic, increasing the flow of urine. The amino acids eventually break down to urea, which acts as a diuretic, too. The overall effect is to force the kidneys to work much harder than they should. The nephrons, which are the kidneys' filter units, gradually die off in the process.

We need protein in the diet, but we do not need a large overdose of protein. The problems of calcium loss and kidney damage occur, not just in those who consume high-protein formulas, but in people who consume meat, chicken, or fish on a regular basis.

The best advice about protein is to stick with plant sources. A varied menu of grains, beans, and vegetables contains more than enough protein for human needs. There is no need to carefully combine proteins. Any variety of plant foods provides sufficient protein. When meats are included, the protein content easily becomes more than the body can handle safely. For example, if you were to have a single 7 ounce serving of roast beef, you would get 62 grams of protein. This one serving contains more than the recommended daily allowance of protein for a whole day (a range of 44–56 gms., depending on your age and level of activity), unless you are pregnant or nursing.

Let's take a look at two other high-protein products, egg whites and skim milk. Doctors learned long ago that egg yolks were loaded with cholesterol. A single egg yolk contains 213 mg. of cholesterol (and is 80% fat). That is even more cholesterol than in an 8-ounce steak. But while many doctors now recommend avoiding egg yolks, some still encourage the consumption of egg whites because they contain protein. Well, the fact is that egg whites contain *too much* protein. Of the calories in an egg white, fully 85% are from protein. That is a huge amount that no one needs. (In addition, salmonella bacteria are an increasing problem in eggs, even those with intact shells.)

Protein Myths

1. Milk for strong bones
2. Egg whites are good
3. Vegetarians do not get enough protein

Skim milk is a similar wrong turn. Because of the high saturated fat content of whole milk, many people have chosen skim dairy products. Getting rid of the dairy fat is certainly a good idea, because the fat in whole milk, butter, cheese, cream, and ice cream will tend to increase cholesterol levels and elevate cancer risk. But after the fat is removed, skim milk is hardly a health food. It contains no fiber, and no complex carbohydrates, but has a substantial amount of lactose sugar (55% of calories). Antibiotics are also frequently present in milk products, due to their routine use on farms.

If you thought you needed milk for strong bones, you have been the victim of an extremely aggressive advertising program by the dairy industry that was not based on good science. The fact is that people in countries that consume milk routinely tend to have weaker bones than those in countries that avoid milk. Osteoporosis is more likely due to excess protein in the diet and to sedentary living. It is not due to a "milk deficiency," and milk consumption does not slow the osteoporosis that commonly occurs in older women.

There are also concerns about the *type* of protein in milk. Milk proteins often cause allergies and other health prob-

41

lems. There are indications that milk proteins contribute to juvenile-onset diabetes, and specialized cow proteins (antibodies) are now known to cause colic in infants. Like it or not, nature designed cow's milk for baby cows, not for people.

LIMIT ALCOHOL

In general, health recommendations have been mixed on alcohol. Modest alcohol consumption—one to two drinks per day—does not promote heart problems. On the other hand, even small amounts of alcohol increase the risk of breast and colon cancer and contribute to birth defects. And, of course, beyond modest use, alcohol contributes to many other very serious health problems, from accidents to heart disease, cancer, neurological disorders, and digestive problems.

What about its effect on your waistline? This is no mystery. Alcohol is fattening. People who consume beer, wine, or mixed drinks on a regular basis get a big load of extra calories, as the list on page 44 shows.

These figures are not presented for you to remember, but rather to illustrate that the alcohol in these beverages packs a significant number of calories.

What is important about alcohol, however, is not just its calorie content. The important point is this: *Alcohol adds to the calories you are already consuming, rather than displacing any*. For example, if you were to eat four bread sticks before dinner, you would eat a bit less at dinner. The 150 calories in the bread would displace about the same amount from the food you would have later. But alcohol does not seem to have this same sort of compensatory mechanism.[7] If you substitute a beer for the bread sticks, it also holds 150 calories, and the calories from alcohol are not compensated for by eating less later. The calories in alcohol *add* to what you eat.

Drink	Calories
Wine (4 oz.)	85
Light Beer (12 oz.)	100
Beer (12 oz.)	150
Liquor (1.5 oz.)	124
(100 proof gin, rum, vodka, or whiskey)	

The presumption is that although soft drinks hold about the same number of calories as beer (a can of cola holds about 155 calories), they are much more likely to be compensated for later than are the calories in alcoholic beverages, since the calories in soft drinks come from sugar. For mixed drinks, the calories in the drink mix are not so likely to contribute to your girth as the calories in the alcohol that is added to it. This is not a recommendation for colas and drink mixes. The point is that alcohol can really widen your waistline. For alcoholics, the effects are different. Alcoholics often consume less food than do non-alcoholics and are deficient in a host of nutrients.

SWEETS AND SWEETENERS

Concentrated sugars, such as hard candies, are just chunks of simple sugars and lack any fiber or water. As a result, they are as concentrated a form of calories as can be found in a carbohydrate food. If you consume large quantities of sugary foods, such as sweets and sodas, you will get more calories than the body needs.

But even so, sugars are not nearly as calorie-dense as fats. If you are not controlling the amount of fat you are eating, there is little point in worrying about sugar.

Often, sugar is not the main problem in sweets. In cookies, pies, and cakes, there may be a lot of sugar, but there is usually a huge amount of fat, too.

Sweets	*Percentage of calories from fat*
Haagen-Dazs ice cream	57%
Hershey dark chocolate bar	50%
Chips Ahoy cookies	42%
Pillsbury German chocolate cake	40%

When selecting sweet foods, pick those with the lowest amount of fat. How about fruit for dessert? And sodas should be replaced with spritzers or water.

Check Your Knowledge

Try each question. Do not skip any. Answers are listed below.

1. How do alcoholic beverages affect weight problems?
2. What problems are caused by diets with too much protein?
3. Is there a lot of fat in pies and cookies?
4. True or false: As far as protein is concerned, the more the better.
5. True or false: Vegetarians get enough protein without carefully combining foods.

Answers

1. Alcohol contains calories, and we do not compensate for this by eating less later.
2. Osteoporosis and kidney problems.
3. Yes.
4. False. We need some protein, and the amount in plant foods is sufficient. Adding high-protein products is not healthful.
5. True.

Forget artificial sweeteners. They are no answer to weight problems. First of all, they do not seem to have much power to help in weight control. Using an artificial sweetener instead of a teaspoon of sugar saves you only 16 calories. But just two grams of fat hold more calories than the teaspoon of sugar. That is not to say that you should consume sugar, but it is to say that artificial sweeteners are a distraction from the real dietary issues, which for most people relate to the fat content of the diet.

More importantly, artificial sweeteners are poisonous. We have seen this over and over again. Cyclamates can cause cancer. The same may be true of saccharin, although it remains on the market with warning labels on each package. Aspartame, marketed under the name Nutra-Sweet, has problems of its own. Substantial evidence links aspartame to a variety of effects on the brain. Headaches are common, and there is currently a scientific debate over whether aspartame can cause grand mal seizures and whether children, including babies developing in the womb, may suffer brain damage if exposed to aspartame. I see no value in chemical sweeteners.

WATCH OUT FOR STUFFING

Most overweight people do not overeat. Most actually eat less than thin people do. But some people do overeat. For one of many reasons, they are "stuffers"; they keep eating long after others would have had enough. It is important to identify whether you are a member of that minority of overweight people who do tend to overeat so you can learn what to do about it. Let's look at three principal reasons for overeating.

1. The Restrained-Eater Phenomenon. As we saw earlier, one big reason for an episode of overeating is the restrained-eater phenomenon, which kicks in after periods of very-low-calorie dieting. This can affect anyone, even people who have never had a tendency to binge or to overeat for any psychological reason. The key, of course, is to avoid the very-low-calorie regimens that tend to produce binges.

2. Eating in Response to Emotions. Ask yourself these questions.

- Is food your usual answer to stress?
- Do you eat when you are not at all hungry?
- Do you eat throughout the day?

If the answer to any of these is yes, then this section may be for you.

Some people eat when they are under stress. Depression, anxiety, hurt feelings, anger, or sadness are answered with a trip to the kitchen. This is easy to identify, and a bit harder to remedy. Do not expect to plumb the depths of your psyche and rearrange its contents in short order. For now, you need to make a plan to compensate for this tendency.

Anticipate that from time to time, like it or not, you will become angry or sad or frustrated with things, and plan to deal with these feelings in another way. Is there someone you can talk to, or someone you can call? If food is serving as a comfort, what other comforts can you take advantage of? For example, are there certain places, photographs, books, or clothes that serve as comforts, too?

Deal with emotions in ways that are inconsistent with eating. For example, if you plan to get together with a friend, be with someone who is not preoccupied with eating, and pick a place where eating will not occur—meet in a park or office instead of a restaurant. Then as meal-time approaches, fill up on healthful foods first.

Some people may use overweight as a defense. A heavy body may fend off intimacy or other anxiety-provoking encounters. The vast majority of overweight people are not in this category, but, if you are, it will be helpful to recognize it.

Are you eating out of boredom? We need many forms of nourishment: friends, intellectual challenges, physical activities, romance, challenges and successes in our lives, rest, and sleep. When these are absent, food may become a cheap substitute. Is food taking the place of something else?

If you are saying, "I overeat, but I do it because food

tastes so good," it may be worth examining what else occupies your time. If your life is filled with boredom, then food may well be the most exciting thing in it. It is important to see what prevents you from engaging more fully in other activities that make life what it is.

Overeaters Anonymous has helped many, many people. Their number is listed in your telephone book. If you really are overeating, OA can be a terrific help.

3. *Carbohydrate Craving.* There is a group of people who have a particular craving for carbohydrates. It is not because of their taste; the foods can be either sweet or starchy. It is apparently due to an effect carbohydrates have on brain chemistry. Carbohydrates boost a brain chemical called serotonin, which is important in brain functions, including sleep and mood regulation. Most antidepressants increase serotonin levels in the brain, among other actions. One theory is that carbohydrate cravers have naturally low levels of serotonin and, so, tend to be depressed. They eat large quantities of carbohydrates because they have noticed that it helps them to feel better.

That is the theory. Here is the chemistry behind it. Carbohydrates break down in the body to sugars, which, in turn, stimulate insulin secretion. Insulin is a hormone produced in the pancreas. It helps get sugar out of the bloodstream and into the cells of the body. Now that is not all insulin does. It also helps amino acids, which are the building blocks of protein, to get out of the bloodstream and into the cells. So, after a carbohydrate-rich meal, insulin drives the sugar and the amino acids out of the blood and into the cells.

Now here is the interesting part: As the insulin drives

the amino acids out of the blood, it leaves behind one particular amino acid called tryptophan. Tryptophan stays behind because it is stuck to a large carrier molecule. Without all the other amino acids around, tryptophan has less competition for getting into the brain. So the tryptophan passes into the brain, where it is converted to serotonin, which can alter moods, and cause sleepiness. The bottom line is that carbohydrate-rich meals increase serotonin in the brain. Carbohydrate cravers tend to become depressed in the winter months when the days are short. Food may help normalize their brain chemistry.

There is nothing wrong with a high-carbohydrate menu. As we have seen, carbohydrates are very important. The key is to select foods rich in complex carbohydrates, such as rice and other grains, beans, and vegetables, rather than sugar candies or sugar-fat mixtures that really will add to one's waistline.

THE ROLE OF PHYSICAL ACTIVITY

Our lives have become all too sedentary. We have elimi-nated most of the physical activities that got our blood moving when we were younger and that kept our ances-tors fit. It is terrific to bring physical activity back into our lives for four reasons.

Movement burns calories

Every movement you make, whether it is blinking your eyes or lifting a grand piano, burns some calories. The more we move, the more calories we burn.

Regular physical activity boosts your metabolism

Calories are burned more quickly, not only while you are exercising, but also afterward for a period of time.

Physical activity helps preserve your muscle mass

Muscle tissue has a rapid metabolism and is much better than fat tissue at burning off the calories we ingest. If your muscles waste away from inactivity, your body burns fewer calories per hour.

Physical activity helps control the appetite

Twenty minutes of exercise before dinner reduces the appetite slightly. This seems to be particularly true for activities that warm the body, such as tennis, running, or dancing. (Some people experience an *increase* in appetite after cooling exercises, such as swimming.) Unfortunately, it is likely that overweight people experience less (or even none) of the exercise-induced change in appetite than do normal-weight individuals, so this may be a mechanism that helps people stay thin rather than helping people to get thin.[8]

There are numerous other benefits of physical activity, from reduced risk of heart disease and cancer to more energy and a more relaxed outlook on life. You may find that you will sleep more soundly when your body is tired from exercise. In turn, better sleep makes you feel like taking care of yourself. Chronically tired people prop themselves up with all sorts of indulgences, including unhealthful foods, that do not seem so important when they are well rested.

HOW MUCH ACTIVITY?

Let's start with a half-hour walk every day, or, if you prefer, an hour three times per week. Pick a place to walk that is enjoyable for you. Enjoy the sights, sounds, and smells.

Feel free to substitute any equivalent activity in place of walking. Here are some examples of physical activities and the number of calories they burn.

Activity	Calories
A brisk half-hour walk	120
A leisurely half-hour bicycle ride	140
A half-hour ping-pong game	210
A half-hour swim	240
A half-hour jog	284
An hour of gardening	300
An hour of golf	356
An hour of tennis	456

The key is to have fun. Choose something you'll enjoy. If you like dancing, gardening, bike-riding, a run with your dog, or a vigorous walk in the woods, then off you go! Bring a friend along if you can. Making activity a social event decreases the possibility of drifting back into sedentary living. At work, use the stairs instead of the elevator.

If you have access to a health club, you will find all sorts of sports and physical activities that turn exercise into pleasure. The old gym has really been transformed into an environment that makes physical activity fun and tailors it to the individual.

Start slowly, particularly if you have been sedentary for some time. If you are over forty or have any history of illness or joint problems, talk over your plans with your doctor. And remember, stick with activities that you really enjoy.

If you have been on a low-calorie diet, you should switch to a low-fat, high-carbohydrate menu without calorie restriction before you begin any program of regular vigorous exercise. The reason is that the low-calorie diet probably slowed down your metabolism. Even though exercise will boost the metabolic rate of most people, it can actually have the opposite effect on people who have been starving themselves. So stop the calorie restriction first, then, after a couple of weeks, add physical activity.

WHAT ABOUT GENETICS?

There is one factor we cannot control, and that is our genetic inheritance. Like it or not, if your parents were both thin, you and your siblings will tend to be thin. If your parents were heavy, you will have a similar tendency.

We also tend to inherit our parents' shape. If your parents were apple-shaped, carrying their weight in their chests and abdomens, you are likely to be apple-shaped as well. If they were "pears," carrying their weight in their hips and thighs, you are likely to be pear-shaped as well. There are all sorts of shape variations. Size is more easily changed than shape. If you carry your weight in your hips, as you lose weight you might become a skinny "pear," but you will still be a "pear."

Fat on the abdomen is easier to lose than hip fat. Although hip fat is more difficult to remove, it is also less likely to contribute to health problems. To determine whether you are at greater risk of health problems from being overweight, take a tape measure and measure around your waist and around your hips. For men, increased risk of health problems begins when your waist is bigger than your hips. For women, they begin when your waist is more than 80 percent of your hip measurement.

If you have passed that point, your weight problem is not a cosmetic issue anymore. It is a very real contributor to heart problems, cancer, diabetes, and a broad range of other problems that you do not want to have.

Some people believe that, because there is a strong genetic component to our size and shape, there is nothing they can do to lose weight. This is not true. Although the genetic factors that are passed from parents to children exert important effects, we do not just give our children DNA. We also give them recipes. We give them attitudes about food and preferences for various kinds of food. We also tend to pass along an interest or disinterest in physical activity, and attitudes about health and about how our bodies should look. These can all be modified, if we decide to do so. Whatever hand we have been dealt by our inheritance, there are still steps we can take to change our weight.

The key about genetics is to remember that it is only one of several factors that affect your weight. It shows your tendency. But within that tendency, there is a great deal that you can do to help reach the body size you want.

A WORD ABOUT B$_{12}$

Vitamin B$_{12}$ is needed in small amounts for healthy blood and healthy nerves. It is not made by plants or animals; it is made by microorganisms, such as bacteria and algae. In traditional societies, B$_{12}$ is produced by bacteria found in the soil and on vegetables. It also can occur naturally in the process of preparation of foods, such as Asian miso or tempeh, soy foods which are loaded with the vitamin. In the West, modern hygiene and pasteurization have eliminated these traditional sources. Meat-eaters get the vitamin B$_{12}$ that bacteria produce in the digestive tracts of animals and that passes into the animals' tissues, but people who adhere to a vegetarian diet (as I recommend) should pick up a supplement.

All common multivitamins (One-A-Day, Flintstones, StressTabs, etc.) contain B$_{12}$. The RDA for B$_{12}$ is only 2 micrograms per day. Health food stores carry vegetarian vitamin brands which are free of dairy and meat extracts. Your body has a very good supply of this vitamin already, but if you follow a vegetarian diet, you should begin taking at least 5 micrograms per day of any common supplement of vitamin B$_{12}$.

SUMMARY OF BASIC CONCEPTS

Let's summarize the major points.

Diet: The overall dietary changes are simple.
- Eat foods from plant sources: grains, beans, vegetables, and fruits.
- Avoid animal products.
- Keep vegetable oil to a minimum as well.

What these simple steps do is to cut way down on fat, reduce protein content moderately, and give us the metabolic boost of carbohydrates, plus lots of fiber.

- Avoid calorie restrictions.

Unless you are really stuffing yourself, you can enjoy unlimited quantities of foods. If you really are overeating, you will need to address the psychological factors that prevent you from treating your body better.

- Refined sugars and alcohol should be avoided as well.

It is not necessary to resolve to change your eating habits for the rest of your life. All you can change is what you are doing today. And tomorrow, you can make the same decision again, if you like. But you do not need to plan what you will eat twenty years from now. I point this out because sometimes the idea of lifelong change can be frightening. Don't worry about it. *All you need to work on is what you are doing today. And if you like it, you can stick with it.*

To get the results you want, do not water down these guidelines. Adding occasional servings of chicken or french fries will erode your progress. Give yourself the best.

Physical Activity:

- Walk for a half-hour per day or an hour three times a week.

Or substitute any equivalent activity. Have fun. The cumulative effect can be enormous.

PART II

Let's Get Started

LET'S GET STARTED

A Step-by-Step Program

In Part I, we learned the basic concepts. Now it's time to put them into action by tasting new foods, trying new food stores, activating our muscles, and making a solid change in our lives.

Adapting to a new, improved menu is surprisingly easy. It does take a little time, about three to six weeks for a new habit to become routine, but soon you will wonder why you did not try this before.

First, give yourself a pat on the back. You wanted a change for the better, and here you are, well on your way there. You've already covered a lot of ground in learning about slimming down. Now we will put knowledge into practice. Here is how we will proceed.

1. We will throw out all the high-fat foods that have caused so many problems for us.
2. We will bring in new and interesting foods that are powerful for permanent weight control.
3. We will set up a simple, effective program of

physical activity that will help burn off the pounds.

PLANNING FOR SUCCESS

One reason this program works is that it provides a very powerful food program, linked with physical activity. But it does more: it takes into account what human beings need in order to make a major change in eating habits.

Let's face it: it's not always easy to change habits. *But certain things make it much easier and much more likely to stick.*

I recently reviewed several major research projects in which people were asked to change their diets. In some of these projects, the participants changed their diets very dramatically, and in others they hardly changed at all. It became very clear that the programs that yielded the most changes were those which had certain things going for them. These factors have been incorporated in this program.

1. *Asking for the degree of change you want.* Doctors or researchers who sell their patients short with weak recommendations get nothing more than they ask for.
2. *Don't just read about foods; taste them as well.* That will be a major focus of this chapter.

3. *Go for maximal reward.* Nothing is more encouraging than success. So this program not only uses foods, it also brings in enjoyable physical activity designed for what people want most: permanent weight control.

4. *Simplicity in foods.* This program is designed to be easy to remember, with no need for charts, measuring, or limiting serving sizes. The two rules of thumb are:

   • Use no animal products.
   • Keep vegetable oils to an absolute minimum.

   These simple guidelines are extremely powerful for long-term weight control.

5. *Foods must be enjoyable.* That means quality and quantity. So this program will use tasty foods and no calorie restriction.

6. *Change completely.* Do not tease yourself with foods. Anyone who has tried to change a habit knows about this one. Let's take smoking as an example. If people try only to cut down on smoking, they get essentially nowhere, because the taste for tobacco is always fresh in their minds. It is very easy to increase the frequency of a habit that has not yet been broken. But if they quit completely, they can get some distance between themselves and tobacco and can start a new habit—the habit of being a non-smoker. The same is true of foods. If you have fried chicken or potato chips once a week, then you are con-

stantly teasing yourself with the taste of these very fattening products. If, however, you get away from these foods completely, you allow a new habit to start; you are getting the force of habit working with you, starting a new habit.

7. *Think short-term.* There is no need to make any resolution about what you will do in the distant future. All you can control is what you are doing now. So plan to follow this program for three weeks. At the end of that time, see how you feel. Notice its effect on your waistline. And if you like what you see, you can try it again for another twenty-one days. If you continue, you will get its full benefit. If you stop, you will lose all your gains. But think short-term; do not burden yourself with feeling the need to make resolutions for the distant future.

8. *Family and friends.* Our families eat with us. They eat the food we prepare or, perhaps, prepare the food we eat. Having them on our side is a terrific boost. When researchers have worked with patients to modify their diets, they have found that including the family makes a tremendous amount of difference. So ask them to join you in this program. Now, they may not feel a need for any permanent change in their eating habits. And you do not need to ask them to change permanently. All you need to ask them is to join you while you are working on this program. In many cases, they will want to read it with you. That is the ideal. They will benefit from this program,

too. This new way of eating not only slims waist-lines; it can also lower cholesterol levels, help control blood pressure, help prevent cancer, and prevent many conditions from constipation to varicose veins. At the very least, however, your family and friends must not tempt you with un-healthful foods while you are working through these lessons, and you must not prepare any un-healthful foods for them.

Frequently families get stuck in old habits. They may even want to talk you out of changing and may forecast failure. In that case, you need to have a short sitdown talk. Tell them that if they care about you, they will understand that this is very important to you. They will help and not hinder you. If you have done this with suf-ficient sincerity, they will be overcome with guilt and will plead for forgiveness. (You might suggest that they can make it up to you by doing your shopping for you.)

Get them to think of it as a three-week ad-venture in eating, including foods from a host of exotic foreign locales. Involve your kids in the kitchen, learning about new foods by helping you prepare them.

Follow this program to the letter. Do not "cheat." You are embarking on a powerful and rewarding program. Give it every chance for maximal success. You deserve no less, and I believe you will be really pleased with the results.

GETTING OFFENDING FOODS
OUT OF THE HOUSE

The first step is to throw out the foods that have been problems in the past and that will get in your way in the future. You can throw them away or give them away, but the key is to get them completely out of the house.

Before you begin, have a meal. It is very difficult for a hungry person to throw any food product away, no matter how unhealthful it may be. Then go on a search and destroy mission for the high-fat, no-fiber foods in your house. Get rid of all of the following:

Any meat, poultry, or fish products
All dairy products, including butter, milk or cream, yogurt, and cheese
Margarine
Vegetable oil (Yes, even olive oil.)
All salad dressings other than non-fat dressings
Cookies, cakes, pies, and ice cream, other than non-fat

Potato chips
Nuts and nut butters
Sugary candies

You may notice a certain sense of relief as you rid yourself of these unhealthful products. Now we will stock our shelves with foods that will help us have the body we want.

GETTING TO KNOW NEW FOODS

We are going to go shopping for new kinds of foods and will learn how to prepare them. Don't worry, they are all easy to fix. Our goal for right now is not to be gourmet cooks, but to learn new kinds of foods and to adapt our tastes to a lower-fat menu. If you like, gourmet recipes can come later.

The foods listed on the following pages create a simple beginning menu, but they are important for you to get to know. They are powerful for slimming down. Some are already quite familiar; others may be new. Some will seem very humble, but do not be deceived. They taste terrific and also have an excellent combination of nutritional factors that make them among the most powerful foods for keeping pounds off permanently.

> • Eat before shopping

The emphasis in this list is on convenience and simplicity, with minimal preparation time. For example, you will see canned beans, rather than dried beans. Later, you may wish to cook from scratch, instead of using canned varieties. Frozen vegetables are also included for convenience. Their nutritional value is generally good, and better than canned vegetables.

We will start with a one-week menu. The idea is to stock your kitchen with enough healthful foods to give you a good start on adapting to new food tastes. It will also eliminate the need for frequent shopping trips, which are times of vulnerability to unhealthful impulse purchases.

Most of these items are available from any grocery store, but a few are found only at health food stores, as I will indicate. Notice that you do not have to mix any diet powders, and you certainly do not have to go hungry.

Do not go shopping on an empty stomach. If you do, you risk ending up buying anchovy-packed olives and coconut cream pie and avocado swirl-cake and all the other impulse purchases that seduce hungry stomachs.

People with food allergies should obviously skip any food to which they are allergic. If you are on a sodium-restricted diet, look for low-sodium varieties of canned foods, or compensate with less sodium in other foods. Health food stores often stock low-sodium products.

People on prescribed diets should follow this program in consultation with their health care professional. For example, because these foods improve insulin's efficacy, diabetics may need to reduce their insulin use. Similarly, people with high blood pressure may find they need less of their medication as well. Persons with high triglycerides may need to limit fruit.

- Keep in touch with your health care professional

First, let's take a look at the foods we will be starting with. Then I will give you a shopping list you can use to guide your trip to the store.

BREAKFAST

1. **Fresh fruit.** Melon, grapefruit, oranges, bananas, pineapple, or any other fruit you like. It can be your entire breakfast, or just the beginning.
2. **Hot cereal.** Choose from old-fashioned rolled oats, grits, or other hot cereal. Cooked versions are best, but instant is acceptable. If you like, top with cinnamon, or try strawberries, raisins, or other fresh fruit. Do not use milk.
3. **Whole-grain toast.** Have plain, or top with jelly or cinnamon. Do not use butter, margarine, or cream cheese.
4. **Cold cereal with soymilk.** Choose whole-grain cereals. Soymilk is not only free of animal fats, it is also free of the cholesterol and lactose that are found in dairy products. Use only low-fat soymilk, such as Edensoy (soymilk is carried at any health food store, or in regular groceries near the condensed milks).

5. **Black beans on toast.** This Latin American breakfast sounds unusual, but can be popular on both sides of the border. Simply empty a can of black beans into a saucepan and heat. Spoon the beans onto toast and top with a touch of mild salsa or Dijon mustard. One can of black beans holds two generous servings.

6. **Chickpeas (garbanzo beans).** Open a can of chickpeas and rinse. Eat plain or with non-fat salad dressing. One can holds two generous servings.

(If you are like me, you will be surprised to see these last two. But try them! You're in for a treat.)

LUNCH

At lunch time, convenience is often key. Low-fat lunches not only help you slim down, they also prevent the after-lunch fatigue that follows high-fat meals.

1. **Instant soups.** Health food stores stock a wonderful variety of split pea soup, couscous, noodle soups, and others. These are easy to keep in your

desk at work, since you just add hot water. (If you prefer, bring a thermos of vegetable or split pea soup from home.)

2. **Bread, bread sticks, pretzels, Melba toast.**
3. **"Finger vegetables."** Cherry tomatoes, baby carrots, broccoli and cauliflower tops with non-fat dressing.
4. **Fresh fruit.** Enjoy bananas, apples, pears, oranges, etc., as well as fruit like kiwi and fresh pineapple for a change of pace. Avoid avocados.
5. **Sandwich.** Have a CLT: Cucumber, lettuce, and tomato, on whole-grain bread. Add onion and mustard, if you like. Some people like to add sprouts. Unfortunately, most traditional sandwich fillings are loaded with fat. Avoid meats of any kind, cheese, mayonnaise, and peanut butter.
6. **Chickpeas.** It's easy to keep a can of chickpeas in a desk drawer at work. Simply rinse and serve plain, or with non-fat salad dressing.
7. **Leftovers from dinner or breakfast** are always welcome. Microwave if you like.
8. **If you eat lunch in a cafeteria,** enjoy the cooked vegetables, potatoes, or the salad bar, using a twist of lemon juice instead of dressing. Avoid all meats, eggs, dairy products, and keep vegetable oils to an absolute minimum.

DINNER

In planning meals, I suggest starting with vegetables, usually including two different ones at each meal. Then, add a grain or other starch, such as rice, potatoes, or pasta. Include a bean dish and finish with a serving of fruit. Be generous with grains and other starches, and have smaller portions of bean dishes.

1. **Vegetables.** Try broccoli, spinach, carrots, cauliflower, green beans, peas, lima beans, Brussels sprouts, kale, asparagus, succotash, or any other. Any fresh vegetable is fine: celery, Boston lettuce or other greens, plus any other salad vegetables that strike your fancy: peppers, tomatoes, etc. If you like, top vegetables with a sprinkle of lemon or lime juice, minced garlic, onion, or parsley. Avoid butter or margarine, sour cream, and other fatty toppings. Do not use oil as a topping or for sautéing or frying. Frozen vegetables are convenient and similar in nutritional value to fresh vegetables. Choose plain varieties, not those in cream sauce.
2. **Grains and other starches.**
    a. Rice is one of the best foods for slimming down. It is very low in calories and very nutritious. At the grocery store, notice the variety of

boxed rice dishes, such as curried rice, long grain and wild rice, brown rice, pecan rice, basmati rice, risotto with tomato, or rice pilaf. Nearby, you will see the fabulous mixes for couscous, tabouli, and vegetarian burgers. Avoid any mixes with meat products or a high fat content. At the health food store, you will find organic short-grain rice, which is an excellent choice (see recipe on page 161). Or try any other varieties of rice that catch your fancy.

b. Spaghetti with tomato sauce. Whole-grain spaghetti is best. If you are stuck with regular spaghetti, consider it acceptable for now, even though much of its fiber has been removed. Of commercial tomato sauces, choose those lowest in fat.

c. Bread. Whole-grain varieties are always best.

d. Corn. Corn is a grain, not a vegetable. Enjoy the natural taste of corn without butter, margarine, or oil.

e. Potatoes. Baked, mashed (instant is fine), steamed, or boiled. No hash browns, potato chips, or french fries. If you like, a dab of Dijon mustard or ketchup is okay. If you like gravy on potatoes, pick up a can of Franco American Mushroom-Flavor Gravy. Do not add milk to mashed potatoes, and use no butter, sour cream, margarine, cheese, or other fatty toppings.

3. **Legumes (beans, peas, and lentils).**

a. Black beans. Do not skip this one. Black beans are a delightful discovery. They are ex-

tremely low in fat, packed with fiber, and delicious. Top with mild salsa or mustard. If you are
new to black beans, I strongly suggest buying
them canned (note that different brands vary
widely in their sodium content), rather than
cooking up dried beans, which requires considerably more time. In case you were worrying, for
most people, black beans do not seem to cause
much gassiness.

b. Vegetarian baked beans. Several canned
varieties are available at grocery stores, and are
very convenient.

c. Lentil soup. Progresso and other companies
make delicious, low-fat lentil soups. For those on
sodium restrictions, health food stores carry low-
sodium varieties.

4. **Fruits.** Pears, cherries, strawberries, peaches,
apples, bananas, pineapples, and just about any
other fruit make great desserts or garnishes for
other foods.

PART III

Moving into High Gear

MOVING INTO HIGH GEAR

A Step-by-Step Program

Now you have dramatically cut the fat content of your diet, and boosted the carbohydrate and fiber content. This encourages a steady and permanent weight loss, which is improved with exercise. In effect, if your fat stores were a huge bag of water, you just poked a small hole in it. Slowly and surely it will drain out.

In this chapter we will move beyond the simple meals we discussed in Part II, and have a look at how to bring an essentially limitless variety of foods and new food ideas into your life. We will also deal with some of the pitfalls that can sometimes get in our way.

- Slow and steady means permanent weight loss

NEW RECIPES, NEW FOODS

First, let's open the doors to new kinds of foods. The variety of healthful foods is enormous, thanks to the many countries that have had culinary traditions very different from our own and to the ever-growing interest in nutrition.

Eating Out

When choosing restaurants, the best ones for low-fat, vegetarian foods are Chinese, Japanese, and other Asian cuisines, Middle-Eastern, Indian, Mexican, and Italian. Traveling is a time of challenges to those trying to eat in a healthful way, but many fast-food restaurants have responded to the demand with baked potatoes and salad bars. Taco restaurants feature bean burritos, and the turnpikes of the East Coast even feature spaghetti restaurants in their travel plazas. In your car, bring some fresh fruit or sandwiches along. When you book a flight request a vegetarian meal or fruit plate; all airlines now have them.

Collecting New Recipes

The next time you get a chance, browse through the cookbook section of your local bookstore. There is an astounding wealth of new cookbooks, particularly those offering vegetarian cuisine. They are a goldmine of food ideas. Look at the wonderful pastas, Middle-Eastern foods, Asian foods, and on and on. Some use dairy products or more than minimal amounts of oil, but these recipes can often be modified, as noted on pages 85–86. If you have not already seen my earlier book, *The Power of Your Plate*, I would strongly recommend it as a source of in-depth information on a broad range of nutrition issues, as well as many of my own favorite recipes.

New Food Products

Exploring new food products can be delightful as well. Take a look at the health food store. There is a wonderful range of new products, from soups of every variety, dips, and sandwich fillings to exotic entrees that are a snap to prepare. New brands of tortilla chips are baked rather than fried and, so, are oil-free. Many varieties of rice and other grains produced without chemical treatments are there. Delightful beverages are now available ranging from flavored waters to juices and teas.

There is also a huge range of non-meat "transition foods"—foods that take the place of ice cream, mayonnaise, hot dogs, hamburgers, and most other fatty, cholesterol-packed foods. These substitutes are not always good for everyday use, however. While better than the items they imitate, they are often still rather high in fat.

Resource Groups

If there is a vegetarian society in your area, get in touch with them. They have been exploring healthful eating for years, and can be a great source of information.

TUNE UP YOUR MENU

In the last chapter, we brought high-fiber, high-carbohydrate foods to the center of the plate. Now, let's fine tune things a bit.

Whole grains

Let whole grains (as opposed to processed grains) play a greater role in your diet. "Processed" grains are ground (e.g., flour) or have some of the fiber removed (e.g., white rice). Evidence indicates that whole grains release a bit fewer of their calories. So rice, rolled oats, or corn may release even fewer calories than flour made from whole grains. Let rice replace bread.

Raw foods

Be increasingly generous with raw vegetables and fruits. Many people report remarkable weight reductions when they include large amounts of raw fruits and vegetables in their diets. This is partly because these foods are extremely low in fat and high in fiber and carbohydrate, but there may be other contributors that science has not yet unraveled. People with even stubborn weight problems have benefited greatly by eating more raw vegetables and fruits.

When you select raw vegetables, skip the iceberg lettuce; it's mostly water. For a salad, try fresh spinach and the other delightful greens in the produce department. Add peppers, broccoli, celery, carrots, cauliflower, cooked chickpeas, or whatever else you might like. Happily, it is now easier to find vegetables produced without pesticides.

Again, skip dressings that include oil. Enjoy the taste of vegetables without added flavorings.

Modifying recipes

Recipes that are high in fat can often be easily modified to lower fat content. You will note that often the amounts

---

For Fat-Free Frying

If you are sautéing in oil, here's a helpful tip. Let's say your recipe for spaghetti sauce calls for sautéing onions and garlic in olive oil. How many calories are there in three tablespoons of olive oil? Would you believe 360? Instead, try this: put a half-cup of water in a sauce pan and sauté your onions and garlic in simmering water. You get no extra calories and a pleasant, lighter taste.

The new non-stick pans work remarkably well with this method.

---

of oil added to recipes are quite arbitrary. Once you have "reset" your taste for fat, you will automatically want to leave the grease out of the foods you prepare.

Health food stores sell egg replacer that does exactly that, drastically cutting down the fat content of baked goods. An egg-sized piece of tofu will also accomplish much the same thing in a baking recipe. TVP (texturized vegetable protein), a defatted soy product sold at health food stores, replaces ground beef so well that many pizza companies and other ground beef users have already made the switch.

ACTIVITY AND REST

Increase physical activity naturally. If you are comfortable with your current level of physical activity, let the time you give it increase naturally. Do not force it, or it will become a chore. But let yourself enjoy an extra night out dancing or bowling, a day on the golf course, or walking in the park. The keys, again, are *fun* and *frequency.*

Get plenty of sleep. People need rest. Chronically tired people have no energy to exercise. Often they feel so out of sorts that they are tempted to prop up their flagging spirits with food or alcohol. There is no substitute for adequate sleep.

TROUBLESHOOTING

If you have had insufficient results so far, be patient; slow weight loss is more likely to be permanent than rapid weight loss.

But if you did not lose weight at all, then let's review the basics.

- Were you using any oily foods, such as salad dressings, peanut butter, or margarine?
- Were you buying fatty products, such as tofu hot dogs?
- Were there any animal products in your foods?
- How about alcohol?
- Did you miss out on regular physical activity?

If you had problems in any of these areas, now is the time to address the problem squarely. There are always solutions to these problems, and tremendous rewards when you do.

Are you having trouble sticking to healthful foods with friends? It may be that you are looking to others for approval, and afraid that they will not sympathize with your new and healthful way of eating. If so, I have noticed a remarkable thing in the past few years. Every time the subject of vegetarian foods comes up in conversation, or I ask a waiter for a vegetarian entree, such as a vegetable plate or spaghetti when there was not one on the menu, I always find that they already know that this is a very healthful way to eat. Many are already vegetarians themselves, or at least recognize that they should be. So stop worrying.

When eating out with friends, suggest Italian, Chinese, or Mexican restaurants. At American-style restaurants, do not hesitate to ask for a vegetable plate. If you don't see it on the menu, by all means ask. When I am invited to a party, I always say something along the lines of "I am eating vegetarian foods right now, and I don't want to put you to any trouble. How about if I bring along something like a meatless spaghetti sauce or hummus?" Invariably, my offer is declined, because they planned to have a meatless dish or two. Often, they say that their son or husband is a vegetarian, and it is no problem at all.

Some people have digestive troubles. Any major change in diet can be a temporary challenge for the digestive tract. A meat-eater who becomes a vegetarian suddenly has to adapt to a high-fiber diet. If a vegetarian were to become a meat-eater, a similar problem would occur due to the enormous change in the dietary contents. Any change can lead to temporary indigestion or gas. If this happens to you, be aware that the effect is a temporary price you are paying for your past indiscretions.

Some plant foods—certain varieties of beans, in particular—do tend to cause gas. Try to pin down which is the problem food. Pinto beans, for example, may be a problem, while black beans are not. Include more grains in the diet, such as rice, and deemphasize beans.

GOOD LUCK

This program is so elegantly simple, yet it is the most effective way to control your weight permanently. The great part is that there is no need to count calories, skip meals, or eat small portions. You can enjoy food in reasonable quantities, and enjoy it in a slimmer, healthier body.

PCRM can keep you posted on the latest in nutrition through our website: *www.pcrm.org*

Let me wish you the very best of health and success in your new venture.

PART IV

Menus and Recipes

EASY WAYS TO CUT THE FAT

- Get the animal products out of your diet. Meat, dairy products, and eggs are major contributors of fat.
- Grill and oven-roast foods instead of frying them.
- Select baked versions of fried foods, like baked tortilla chips, potato chips, and pretzels.
- Avoid deep-fried foods like doughnuts; have a bagel instead.
- Use nonstick pots and pans, which allow food to be prepared with little or no added fat.
- "Sauté" in liquid instead of in fat. Heat ½ cup of liquid (water, vegetable stock, wine, dry sherry) in a large pot or skillet. Add the vegetables to be sautéed and cook over medium-high heat, stirring occasionally, until the vegetables are tender.
- When it is absolutely necessary to sauté or fry in oil, nonstick vegetable oil sprays allow you to do so with a fraction of the fat.
- Use a microwave oven to reheat foods.
- Use fat-free dressings on salads.

- Seasoned rice vinegar and balsamic vinegar are great for dressing salads.
- Eat cooked vegetables plain, or with a sprinkling of seasoned rice vinegar or lemon juice.
- Try your favorite fat-free salad dressing on cooked vegetables.
- Replace the oil in salad dressing recipes with seasoned rice vinegar, vegetable stock, bean cooking liquid, or with water.
- Use this cornstarch and water mixture to thicken salad dressing. Whisk 1 tablespoon of cornstarch with 1 cup water. Heat in a small saucepan, stirring constantly, until thick and clear. Refrigerate. Use in place of oil in any salad dressing recipe. May be kept refrigerated for up to three weeks.
- Make soup thick and creamy by adding a cooked mashed potato or instant mashed potato flakes.
- Limit your consumption of avocados, nuts and seeds, and coconut, which are all high in fat.
- Replace ice cream with fruit sorbet or a frozen fruit smoothie.
- The amount of fat in baked goods can often be reduced without affecting the taste or texture of the finished product.
- Experiment with your recipes: start by reducing the fat by half. You may have to add a bit of extra liquid to achieve the desired consistency.
- Applesauce, mashed banana, prune puree, or canned pumpkin may be substituted for the fat in many baked goods.

A GLOSSARY OF INGREDIENTS
THAT MAY BE NEW TO YOU

Most of the ingredients in the recipes are commonly available in grocery stores. A few that may be unfamiliar are described below.

Balsamic vinegar is a reddish-brown vinegar from Italy. It has a robust sweet-sour taste.

Basmati rice is a flavorful variety of long-grain rice sold in natural food stores and many supermarkets. Brown basmati rice is less refined than white basmati rice.

Bulgur is cracked whole wheat that has been parboiled and dried. It is then ground and comes in fine, medium, or coarse grind. Use medium or coarse for a more interesting texture. Bulgur is a pleasant alternative to rice or pasta. It is the key ingredient to the Middle Eastern salad called tabouli.

Cilantro is an herb—the fresh leaves from the coriander plant. Its distinctive flavor is frequently found in Indian, Mexican, and Chinese cooking. It is sometimes called Chinese parsley.

Couscous is made from wheat that is cracked rather than ground. It is sold in the grain section of many supermarkets, natural food stores, and ethnic markets. Whole wheat varieties are found in some natural food stores.

Diced chilies are mild chilies (like Anaheim), which are available canned or fresh. Ortega is a widely sold brand. If you use fresh chilies, remove the skin by charring it under a broiler then rubbing it off.

Eggless mayonnaise is just that—a mayonnaise made without eggs, which is often lower in fat. A good brand available in most health food stores is called Nayonaise; it contains 3 grams of fat per tablespoon compared to regular mayonnaise, which has 11 grams of fat.

Nutritional yeast adds flavor and nutritional value to foods, and is a rich source of B vitamins (look for brands that include vitamin B_{12}). It is similar to brewer's yeast, but better tasting. You'll find it in natural food stores.

Red pepper flakes are dried, crushed chili peppers sold in the Mexican food or spice section of many supermarkets.

Reduced-fat tofu is found in natural food stores and many supermarkets. Also try silken tofu, a smooth, delicate variety that is excellent for sauces, cream soups, and dips. One popular brand, Mori-Nu, is available in most grocery stores.

Rice milk is a mild-flavored beverage made from rice. Delicious on cereal and as replacement for cow's milk in most recipes. Available in natural food stores and a growing number of supermarkets.

Roasted red bell peppers add great flavor and color to dishes. Roast your own or purchase them already roasted, packed in water, in most grocery stores.

Seasoned rice vinegar is flavored with a bit of sugar and salt and is great for salad dressings and on cooked vegetables. Available in most grocery stores.

Soymilk has none of the cholesterol or animal fat of cow's milk. It comes in a variety of flavors, as well as in low-fat and vitamin- and mineral-fortified versions. Soymilks vary widely in flavor, so try several to find the ones you like best. Available in natural food stores and in many supermarkets.

Swanson's Vegetable Broth is widely available and very convenient.

Tahini is sesame seed butter used in Middle Eastern cooking. It is available in natural food stores and many supermarkets.

TVP or textured vegetable protein is made from soybeans. It is a fat-free meat substitute that easily absorbs flavorful sauces. Available dried in small, flake-like pieces that when reconstituted take the place of ground beef in spaghetti sauce, sloppy joes, tacos, etc.

Tofu, sometimes called bean curd, is also made from soybeans and has a soft, cheeselike texture. It is an extremely versatile ingredient that takes on the flavor of any sauce. The soft silken tofu is especially smooth and when pureed is a good substitute for cream.

Unbleached flour is white flour that has not been chemically whitened. Available in most grocery stores.

Whole wheat pastry flour is milled from soft spring wheat. It retains the bran and germ and, at the same time, produces lighter-textured baked goods than regular whole wheat flour. Available in natural food stores.

SEVEN DAYS OF GREAT MEALS

DAY ONE

Breakfast
Orange Julius (page 109)

Whole Grain Toast

Unsweetened Fruit Preserves

Lunch
Simple Spanish Rice (page 165)

Quick Black Bean Chili (page 198)

South of the Border Salad (page 137)

Dinner
Sweet and Sour Vegetable Stir-fry (page 184)

Brown Rice (page 161)

Steamed Kale (page 175)

Peach Cobbler (page 208)

DAY TWO

Breakfast

Whole Wheat Corncakes (page 113)

Fresh Fruit Slices

Lunch

Mixed Greens Salad (page 136)

Balsamic Vinaigrette (page 133)

Potato and Corn Chowder (page 148)

Whole Grain Bread

Dinner

Broccoli Burritos (page 196)

Bulgur (page 166)

South of the Border Salad (page 137)

Melon Slices

DAY THREE

Breakfast

    Cinnamon Orange French Toast (page 118)

    Maple Syrup

    Fresh Fruit

Lunch

    Tomato Soup (page 146)

    Mixed Greens Salad (page 136)

    Simple Piquant Dressing (page 132)

    Whole Grain Bread

Dinner

    Pita Wedges with Hummus (page 224)

    Curried Mushrooms and Chickpeas (page 204)

    Couscous (page 168)

    Steamed Kale (page 175)

    Dried Figs

DAY FOUR

Breakfast

Quick Breakfast Rice (page 122)

Fresh Fruit Smoothie (page 111)

Lunch

Ensalada de Frijoles (page 144)

Warm Corn Tortillas

Dinner

Polenta with Hearty Barbecue Sauce (page 182)

Mixed Greens Salad (page 136)

Simple Piquant Dressing (page 132)

Potatoes with Kale and Corn (page 176)

Chocolate Torte (page 210)

DAY FIVE

Breakfast

Fresh Fruit & Bagel Breakfast (page 120)

Fresh Fruit Smoothie (page 111)

Lunch

Hoppin' John Salad (page 143)

Curried Sweet Potato Soup (page 157)

Whole Grain Bread

Dinner

Thai Vegetables with Rice (page 188)

Braised Cabbage (page 173)

Ginger Peachy Bread Pudding (page 206)

DAY SIX

Breakfast
Blueberry Muffins (page 125)

Hot Cereal

Fresh Fruit

Lunch
Golden Mushroom Soup (page 150)

Cucumbers with Creamy Dill Dressing (page 135)

Garlic Bread (page 129)

Dinner
Vegetarian Burger (page 192)

Potato Salad (page 139)

Corn on the Cob

Watermelon

DAY SEVEN

Breakfast

    Cinnamon Apple Oatmeal (page 121)

    Fresh Peach Smoothie (page 112)

Lunch

    Minestrone (page 152)

    Three Bean Salad (page 141)

    Garlic Bread (page 129)

Dinner

    Chili Mac (page 190)

    Mixed Greens Salad (page 136)

    Balsamic Vinaigrette (page 133)

    French Bread

    Banana Freeze (page 259)

STOCKING YOUR PANTRY
FOR HEALTHY EATING

By keeping some basic foods on hand, you can prepare quick, nutritious meals at a moment's notice. Some of the items listed below are ingredients that show up frequently in recipes.

Others are healthful, quick choices for those days when there just isn't time to cook. If you were to take an inventory of my pantry, here is what you would find (brand names for some items are shown in parentheses)

Dry Goods
cold breakfast cereals without added fat or sugars
hot breakfast cereals
eggless pasta: fettuccine, spirals, lasagne, etc.
brown basmati rice
white basmati rice
quick cooking brown rice (Lundberg)
bulgur wheat
couscous
polenta

rolled oats
whole wheat flour
whole wheat pastry flour
unbleached flour
cornmeal
popcorn
dried lentils
split peas
pinto beans
raisins
instant soups: vegetarian soup cups, ramen soups
instant black beans (Taste Adventure, Fantastic World Foods)
instant pinto beans (Taste Adventure, Fantastic World Foods)
reduced-fat silken tofu (Mori-Nu)
vegetable broth: powder, cubes, or canned
vegetable oil spray
baking soda
baking powder

Canned Foods (canned goods should be used within one year)

canned beans, including kidney, garbanzo, black, pinto, etc.
canned tomatoes, tomato sauce, tomato paste
canned pumpkin
applesauce
vegetarian soups (made without fat)
vegetarian baked beans (Bush's, Heinz)

vegetarian chili beans (Dennison's, Health Valley)
fat-free refried beans (Rosarita, Old El Paso, Bear-
itos)
vegetarian spaghetti sauce (made without fat)
salsa

Breads
whole grain bread (may be frozen)
flour tortillas (may be frozen)
corn tortillas (may be frozen)

Produce
pre-washed salad mix
pre-washed spinach
broccoli
kale or collard greens
yellow onions
garlic
red potatoes
russet potatoes
yams or sweet potatoes
winter squash
green cabbage
carrots
celery
apples
oranges
bananas

Frozen Foods (should be used within 6 months)
apple juice concentrate
orange juice concentrate
frozen corn
frozen peas
frozen bananas
frozen berries
frozen chopped onions
frozen diced bell peppers
fat-free vegetarian hot dogs (Smart Dogs, Yves Veggie Wieners)
fat-free vegetarian burgers (Gardenburger)

Seasonings and Condiments
herbs and spices: cinnamon, ginger, cloves, ground cumin, cayenne, chili powder, red pepper flakes, curry powder, basil

BREAKFAST FOODS

ORANGE JULIUS

Makes 2 one-cup servings

If you ever enjoyed an Orange Julius at a shopping mall or the county fair you're in for a real treat, because this one is as good for you as it tastes. It's quick to make too, if you keep frozen bananas on hand. Just peel the bananas, break them into chunks, and pack them loosely in an airtight container to freeze. They will keep well for about 3 months.

1 large orange, peeled
¼ cup vanilla soymilk or rice milk
¼ cup oat bran
1 teaspoon sugar

½ teaspoon vanilla
1½ cups frozen banana chunks

Combine all ingredients in a blender and process on high speed until completely smooth and thick.

Per 1 cup serving: 170 calories; 4 grams protein; 35 grams carbohydrate; 1 gram fat; 13 mg sodium; 0 mg cholesterol

FRESH FRUIT SMOOTHIE

Makes 2 servings

Start your day out right with this delicious and satisfying beverage.

    1 large orange, peeled
    1 cup frozen banana chunks
    1 cup frozen strawberries
    ½ cup apple juice

Combine all ingredients in a blender and process on high speed until completely smooth.

Per 1 cup serving: 134 calories; 2 grams protein; 30 grams carbohydrate; 0 grams fat; 6 mg sodium; 0 mg cholesterol

FRESH PEACH SMOOTHIE

Makes 1 serving

This smoothie is so thick and rich it's actually like a frozen dessert. Keep frozen bananas on hand so you can whip it up any time.

    1 fresh ripe peach (or nectarine)
    1 frozen banana (about 1 cup banana chunks)
    ½ cup vanilla soymilk

Combine all the ingredients in a blender and process on high until completely smooth.

Per serving: 187 calories; 4 grams protein; 40 grams carbohydrate; 1 gram fat; 46 mg sodium; 0 mg cholesterol

WHOLE WHEAT CORNCAKES

Makes 16 three-inch pancakes

Serve these nutritious pancakes with fresh fruit, fruit preserves, or maple syrup.

½ cup whole wheat flour
½ cup cornmeal
½ teaspoon baking powder
¼ teaspoon baking soda
¼ teaspoon salt (optional)
1 small ripe banana, mashed
2 tablespoons brown sugar
1 tablespoon vinegar
1–1¼ cups soymilk or rice milk

Stir the flour, cornmeal, baking powder, baking soda, and salt together in a mixing bowl.

In a separate bowl, combine mashed banana, brown sugar, vinegar, and 1 cup of soymilk or rice milk. Add to the flour mixture and stir just enough to remove lumps and make a pourable batter. Add a bit more milk if the mixture seems too thick.

Preheat a nonstick skillet or griddle. Spray lightly with a

vegetable oil spray. Pour small amounts of batter onto the heated surface and cook until the tops bubble. Turn with a spatula and cook the second side until golden brown, about 1 minute. Serve immediately.

Per pancake: 44 calories; 1 gram protein; 9 grams carbohydrate; 0 grams fat; 65 mg sodium; 0 mg cholesterol

BUCKWHEAT PANCAKES

Makes about 20 three-inch pancakes

These pancakes are made with yeast, which makes them light and tender without added fat. Start the batter about 1½ hours before you want to serve the pancakes, since it needs an hour to rise.

  1 cup buckwheat flour
  ½ cup unbleached or whole wheat pastry flour
  ½ package active dry yeast (1½ teaspoons)
  ½ teaspoon salt (optional)
  1¼ cups very warm water (about 100°F)
  ½ teaspoon baking soda
  1 tablespoon molasses
  ¼ cup very warm water (about 100°F)

Preheat the oven to 200°, then turn it off. In a large bowl, mix the buckwheat and unbleached flours, the yeast, and salt.

Stir in 1¼ cups of warm water and beat until smooth. Cover the bowl with a plate and place it into the preheated oven to rise until very bubbly, about 1 hour (be sure the oven is turned off!).

Dissolve the soda and molasses in ¼ cup of very warm water and add it to the buckwheat batter. Heat a nonstick skillet or griddle. Mist it lightly with vegetable oil spray. Pour small amounts of batter onto the heated surface and cook until puffed and bubbly. Turn and cook the second side about 1 minute. Serve immediately with syrup, fruit preserves, or fresh fruit.

Per pancake: 31 calories; 1 gram protein; 6 grams carbohydrate; 0 grams fat; 75 mg sodium; 0 mg cholesterol

SOURDOUGH WAFFLES

Makes 8 five-inch waffles

Sourdough starter can be used to make delicious, light waffles without any added fat or cholesterol. Specialty food stores sell sourdough starter mixes, or better yet, get some from a friend who keeps sourdough starter. The care and feeding of sourdough is really quite simple. Simply use it every two weeks and replenish it with equal amounts of flour and water.

   1 teaspoon baking soda
   2 tablespoons maple syrup
   2 cups sourdough starter

Combine the soda and maple syrup in a mixing bowl and stir to mix. Add the sourdough starter and mix well.

   Cook the batter in a preheated, oil-sprayed waffle iron until golden brown, 3 to 5 minutes. Serve immediately.

Per waffle (without oil): 58 calories; 2 grams protein; 13 grams carbohydrate; 0 grams fat; 103 mg sodium; 0 mg cholesterol

CINNAMON ORANGE
FRENCH TOAST

Makes 4–6 slices

French toast is quick to prepare, low-fat, and cholesterol-free when you make it without eggs. Vary the bread you use for different flavors and textures (cinnamon raisin is my favorite).

1 cup soymilk or rice milk
3 tablespoons cornstarch
1 tablespoon orange juice concentrate
1 tablespoon maple syrup
1 teaspoon vanilla
½ teaspoon cinnamon
⅛ teaspoon salt
4–6 slices whole grain bread

Mix the soymilk or rice milk, cornstarch, orange juice concentrate, maple syrup, vanilla, cinnamon, and salt. Pour the mixture into a broad, flat dish. Coat each slice of bread on both sides with the mixture.

Preheat a nonstick skillet or griddle and mist it with a

vegetable oil spray. Add the bread slices and cook until the first side is golden brown, about 3 minutes. Turn carefully with a spatula and cook the second side until golden brown, about 3 minutes. Serve hot with maple syrup or fruit preserves.

Per slice: 119 calories; 3 grams protein; 24 grams carbohydrate; 1 gram fat; 143 mg sodium; 0 mg cholesterol

FRESH FRUIT & BAGEL BREAKFAST

Serves 1

Enjoy a bagel and fresh fruit for a quick and satisfying breakfast or snack.

  1 bagel, any flavor (except cheese, egg, or chocolate chip!)
  sliced fresh fruit (banana, orange, strawberries, peach, etc.)

Slice the bagel in half and toast it lightly. Place a layer (or several layers) of freshly sliced fruit on one half of the bagel and top it with the remaining half. Enjoy!

Per bagel: 265 calories; 7 grams protein; 52 grams carbohydrate; 1 gram fat; 525 mg sodium; 0 mg cholesterol

CINNAMON APPLE OATMEAL

Makes 3 cups

1 cup old fashioned rolled oats
1 green apple, cored and diced
¼ cup raisins
2 tablespoons chopped dates
¼ teaspoon cinnamon
¼ teaspoon salt
rice or soymilk for serving

Bring 2 cups of water to a boil, then add the oats, apple, raisins, dates, cinnamon, and salt. Bring to a simmer, then cook 3 minutes, stirring frequently. Remove from heat and cover. Allow to sit for 10 minutes. Spoon into bowls and serve with rice milk or soymilk.

Per 1 cup serving: 187 calories; 4 grams protein; 38 grams carbohydrate; 2 grams fat; 181 mg sodium; 0 mg cholesterol

QUICK BREAKFAST RICE

Makes about 4 cups

1½ cups vanilla soymilk
1 tablespoon cornstarch
2 cups cooked brown rice
¼ cup maple syrup
⅓ cup raisins
¼ teaspoon cinnamon
1 teaspoon vanilla

Pour the vanilla soymilk into a medium-sized saucepan. Add the cornstarch and whisk until smooth. Stir in the rice, maple syrup, raisins, and cinnamon. Bring to a simmer over medium heat.

Cook 3 minutes, then remove from heat and stir in the vanilla.

Serve hot or cold.

Per 1 cup serving: 233 calories; 4 grams protein; 51 grams carbo-hydrate; 1 gram fat; 170 mg sodium; 0 mg cholesterol

BREAKFAST CEREALS

Hot and cold cereals can provide a quick and wholesome breakfast if you choose your cereals carefully. When shopping for breakfast cereals, the old adage, "The simpler, the better," is helpful to keep in mind. Generally, the fewer ingredients in a breakfast cereal, the more healthful it will be.

When you read the ingredient list, the first ingredient should be a whole grain, or a mixture of whole grains, for example, whole wheat, corn, oats, barley, rice, etc. If the cereal includes salt, make sure it is only a small amount. Check the nutrition label. Ideally, a serving should contain 200 mg or less of sodium per serving.

Many cold cereals contain a lot of sugar, making them essentially a breakfast dessert. Check the ingredients. If sugar appears near the top of the list, or if there are several different forms of sugar included, such as dextrose, corn syrup, honey, barley malt, etc., then the cereal has too much sugar to be truly healthful. The major problem with these cereals is that you tend to eat more than you need, because it's difficult to stop eating sweets.

One other ingredient to watch for in cereals is fat, usually in the form of oil, or partially hydrogenated oil. One of the worst offenders in this category are the granola cereals, which often contain as much fat as many cookies. Cereals containing added fat should be avoided, because they will not help you lose weight. Instead, choose whole grain cereals with as little sugar as possible.

MUFFINS AND BREADS

BLUEBERRY MUFFINS

Makes 12 muffins

1⅔ cups unbleached or whole wheat pastry flour
1½ teaspoons baking soda
½ teaspoon salt
½ teaspoon nutmeg
1 10.5-ounce package firm silken tofu
½ cup sugar
2 tablespoons orange juice concentrate
2 cups blueberries, fresh or frozen

Preheat oven to 350°. Mix the flour, baking soda, salt, and nutmeg in a large bowl.

Place the tofu, sugar, and orange juice concentrate into a

food processor and blend until completely smooth. Add to the flour mixture along with the blueberries and stir until just blended.

Drop spoonfuls of the batter into oil-sprayed muffin cups, and bake in the preheated oven for 25 minutes. Remove muffins from pan and cool on a rack.

Per muffin: 126 calories; 5 grams protein; 25 grams carbohydrate; 1 gram fat; 195 mg sodium; 0 mg cholesterol

BANANA DATE MUFFINS

Makes 12 muffins

2 cups whole wheat pastry flour
2 teaspoons baking soda
½ teaspoon salt
⅓ cup wheat germ
⅓ cup oat bran
4 ripe bananas, mashed (about 2½ cups)
½ cup sugar
¾ cup soymilk or rice milk
1 teaspoon vanilla
⅓ cup chopped dates

Preheat oven to 350°. Mix the flour, baking soda, salt, wheat germ, and oat bran in a large bowl.

In another bowl, mash the bananas and mix in the sugar. Stir in the soymilk and vanilla and mix thoroughly. Add the flour mixture, along with the dates, and stir to mix.

Fill oil-sprayed muffin cups nearly to the top, and bake in preheated oven for 20–30 minutes, until a toothpick inserted into the center comes out clean. Let stand 2 minutes, then remove muffins from pan and cool on a rack.

Per muffin: 174 calories; 5 grams protein; 36 grams carbohydrate; 1 gram fat; 230 mg sodium; 0 mg cholesterol

GARLIC BREAD

Makes one loaf (about 20 slices)

Roasted garlic has a mellow flavor and creamy texture that makes delicious fat-free garlic bread.

2 large heads garlic
1 baguette or loaf of French bread, sliced
1–2 teaspoons mixed Italian herbs (optional)
¼ teaspoon salt

Roast the whole, unpeeled heads of garlic in a 400° oven (or toaster oven) until they feel soft when squeezed. This will take about 30 minutes. Reduce the oven temperature to 350°.

Peel the cloves or squeeze them out of their skin, then mash them into a paste with a fork. Mix in the herbs, if desired, and salt. Spread onto the sliced bread. Wrap tightly in foil and bake for 20 minutes.

Per slice: 91 calories; 3 grams protein; 18.5 grams carbohydrate; 0 grams fat; 179 mg sodium; 0 mg cholesterol

SALAD DRESSINGS AND SALADS

FAT-FREE SALAD DRESSING

Makes ½ cup

Seasoned rice vinegar makes a simple, delicious dressing for salads and cooked vegetables. It will keep in the refrigerator for 2 to 3 weeks.

½ cup seasoned rice vinegar
1–2 teaspoons stoneground or Dijon mustard
1 garlic clove, pressed or crushed

Whisk all the ingredients together. Store in a closed jar or air-tight plastic container.

Per tablespoon: 14 calories; 0 grams protein; 3 grams carbohydrate; 0 grams fat; 310 mg sodium; 0 mg cholesterol

SIMPLE PIQUANT DRESSING

Makes ½ cup

¼ cup of your favorite salsa
¼ cup seasoned rice vinegar

Mix the salsa and seasoned rice vinegar. Refrigerate any extra dressing.

Per 1 tablespoon: 8 calories; 0 grams protein; 7 grams carbohydrate; 0 grams fat; 149 mg sodium; 0 mg cholesterol

BALSAMIC VINAIGRETTE

Makes ⅓ cup

Balsamic vinegar is a delicious wine vinegar from Italy with a mellow flavor that is perfect for salads.

2 tablespoons balsamic vinegar
2 tablespoons seasoned rice vinegar
2 tablespoons water
1–2 garlic cloves, crushed

Whisk all ingredients together.

Per 1 tablespoon: 6 calories; 0 grams protein; 1.5 grams carbohydrate; 0 grams fat; 99 mg sodium; 0 mg cholesterol

CREAMY DILL DRESSING

Makes about 1½ cups

This rich-tasting, creamy dressing has no added oil. It is made with silken tofu, which is available in most markets. Mori-Nu is one popular brand.

1 10.5-ounce package firm silken tofu
1½ teaspoons garlic powder or granules
½ teaspoon dill weed
½ teaspoon salt
2 tablespoons water
1½ tablespoons lemon juice
1 tablespoon seasoned rice vinegar

Combine all the ingredients in a food processor or blender and blend until completely smooth. Store any extra dressing in an airtight container in the refrigerator.

Per 1 tablespoon: 23 calories; 3 grams protein; 2 grams carbohydrate; 0.5 gram fat; 115 mg sodium; 0 mg cholesterol

CUCUMBERS WITH CREAMY DILL DRESSING

Serves 6

1 batch Creamy Dill Dressing (preceding recipe)
2 cucumbers, peeled and thinly sliced
½ cup thinly sliced red onion

Combine all the ingredients in a salad bowl and toss to mix.

Per serving: 64 calories; 6 grams protein; 7 grams carbohydrate; 1 gram fat; 232 mg sodium; 0 mg cholesterol

MIXED GREENS SALAD

Serves 6

For an almost instant salad, use one of the convenient prewashed salad mixes that are available in most markets. Add any other vegetables you enjoy (several are suggested below), and toss with fat-free dressing.

    6 cups prewashed salad mix
    1 tomato, cut into wedges
    ½ red or yellow bell pepper, seeded and sliced
    1 carrot, grated or cut into thin slices
    ½ cup peeled and sliced jicama or cucumber
    3 tablespoons fat-free dressing

Combine all ingredients and toss gently to mix.

Per serving: 20 calories; 0 grams protein; 4 grams carbohydrate; 0 grams fat; 160 mg sodium; 0 mg cholesterol

SOUTH OF THE BORDER SALAD

Serves 6

This salad is a cool and crunchy contrast to Mexican food or any other spicy cuisine. It may be made in advance and keeps well.

1 large carrot, peeled and cut into ¼-inch rounds
1 medium jicama, peeled and diced (about 2 cups)
1 red bell pepper, seeded and diced
1 small sweet onion, thinly sliced (about ½ cup)
2 tablespoons finely chopped cilantro
3 tablespoons Swanson's Vegetable Broth
2 tablespoons seasoned rice vinegar
2 tablespoons lemon juice
1 teaspoon stoneground mustard
¼ teaspoon salt
¼ teaspoon crushed red pepper flakes

Combine the carrot, jicama, bell pepper, onion, and cilantro in a salad bowl. In a small bowl, mix the remaining ingredients.

Pour over the vegetables and toss to mix.

Per serving: 32 calories; 1 gram protein; 7 grams carbohydrate; 0 grams fat; 264 mg sodium; 0 mg cholesterol

POTATO SALAD

Serves 8

You'll love this delicious, old-fashioned potato salad without the added fat or cholesterol. Tofu is the secret for a creamy, nutritious dressing.

6 medium russet potatoes (about 8 cups diced)
5 stalks celery, including leaves, thinly sliced
1 small red onion, finely chopped
1 cup finely chopped parsley
1 cup finely shredded cabbage (optional)
1 10.5-ounce package firm silken tofu
⅓ cup seasoned rice vinegar
¼ cup cider vinegar
2 tablespoons stoneground mustard
1 teaspoon dill weed
1 teaspoon salt
½ teaspoon black pepper
½ teaspoon turmeric

Scrub the potatoes, peel them if you wish, and cut them into cubes. Steam over boiling water until just tender, about 15 minutes, then transfer to a large bowl. Add the celery, onion, parsley, and cabbage.

In a blender, combine all the remaining ingredients. Blend until completely smooth, then pour over the salad. Toss gently to mix. Chill if time allows.

Per serving: 184 calories; 6 grams protein; 38 grams carbohydrate; 1 gram fat; 461 mg sodium; 0 mg cholesterol

Tip: If the salad is kept for a while, the potatoes will absorb some of the dressing. You may wish to add a tablespoon or two of water to restore a moist, creamy texture.

THREE BEAN SALAD

Serves 8

This salad is delicious all by itself or as an addition to a green salad.

1 15-ounce can kidney beans, drained
1 15-ounce can garbanzo beans, drained
1 15-ounce can green beans, drained
½ small red onion, finely chopped
4 tablespoons finely chopped fresh parsley
½ cup cider vinegar
2 tablespoons seasoned rice vinegar
3 garlic cloves, minced
½ teaspoon basil
¼ teaspoon each: oregano, marjoram, and black pepper

Drain the beans and toss them with the chopped onion and parsley.

In a separate bowl whisk the remaining ingredients together, then add to beans and toss to mix. Refrigerate and allow to marinate 2 to 3 hours before serving.

Per serving: 141 calories; 7 grams protein; 26 grams carbohydrate; 1 gram fat; 140 mg of sodium; 0 mg cholesterol

HOPPIN' JOHN SALAD

Makes about 6 cups

Black-eyed peas give this salad a wonderfully distinct flavor.

2 15-ounce cans black-eyed peas, drained
2 cups cooked brown rice
2 green onions, sliced
1–2 garlic cloves, crushed
1 stalk celery, sliced (about ½ cup)
2 small tomatoes, diced
2 tablespoons finely chopped parsley
⅓ cup lemon juice
½ teaspoon salt

Combine the black-eyed peas, cooked rice, green onions, garlic, celery, tomatoes, and chopped parsley in a salad bowl.

Mix the lemon juice and salt, then pour over the salad and toss gently to mix. Chill 1 to 2 hours if time permits.

Per serving: 110 calories; 5 grams protein; 21 grams carbohydrate; 0.5 gram fat; 280 mg sodium; 0 mg cholesterol

ENSALADA DE FRIJOLES

Serves 4

This recipe makes four very generous salads that are substantial enough to be a complete meal. The avocado adds wonderful flavor, a bit of fat, and vitamin E. Leave it out if you prefer.

2–3 cups cooked brown rice (page 161)
8 cups salad mix
2 carrots, grated or cut in thin strips
1 15-ounce can black beans, drained
1 cup diced jicama
2 tomatoes, sliced
1 15-ounce can corn, drained
½ cup cilantro leaves
½ avocado, thinly sliced (optional)
¼ cup of your favorite salsa
¼ cup seasoned rice vinegar
1 garlic clove, crushed or pressed
additional salsa for topping

Heat the brown rice and make a bed of it on each of four plates. Top with a layer of salad mix and grated carrot.

Rinse the beans, then sprinkle them over each of the salads. Add the jicama, tomatoes, corn, cilantro, and avocado. Mix the salsa, seasoned rice vinegar, and crushed garlic. Sprinkle over the salad, then top with a generous spoonful of salsa.

Per serving (with avocado): 344 calories; 11 grams protein; 63 grams carbohydrate; 5 grams fat; 357 mg sodium; 0 mg cholesterol

Per serving (without avocado): 302 calories; 10 grams protein; 60 grams carbohydrate; 2 grams fat; 355 mg sodium; 0 mg cholesterol

SOUPS

TOMATO SOUP

Serves 6 to 8

Hot, steamy tomato soup is one of everyone's favorite foods. This quick version is made with canned tomatoes and takes just 20 minutes to prepare.

1 small onion, chopped
3 stalks celery, sliced
1 28-ounce can crushed tomatoes, including liquid
2½ teaspoons sugar
½ teaspoon paprika
½ teaspoon basil
¼ teaspoon pepper
3 cups soymilk

1 cup cooked brown rice
½ teaspoon salt

Combine the onion, celery, tomatoes, sugar, paprika, basil, and pepper in a pot. Cover and simmer 15 minutes, stirring occasionally. Transfer 3 cups of the soup to a blender. Add the soymilk and blend until smooth. Return to the pot and add the rice and salt. Heat over a low flame until very hot and steamy.

Per serving: 137 calories; 4 grams protein; 28 grams carbohydrate; 1 gram fat; 265 mg sodium; 0 mg cholesterol

POTATO AND CORN CHOWDER

Serves 8

This is a mildly spicy chowder that is delicious with corn-bread and a green salad with Simple Piquant Dressing (page 132).

4 russet potatoes, peeled and diced (about 4 cups)
2½ cups water or vegetable stock
1 yellow onion, chopped
2 garlic cloves, minced
1 red bell pepper, diced
1 teaspoon ground cumin
1 teaspoon basil
1 teaspoon salt
¼ teaspoon black pepper
1 4-ounce can diced chilies
2 cups corn, fresh or frozen
1–2 cups soymilk

Combine the diced potatoes in a pot with 2 cups of water or vegetable stock. Cover and cook until tender, about 20 minutes.

Heat the remaining ½ cup of water or stock in a large

pot and cook the onion, garlic, and bell pepper for 3 minutes. Add the cumin, basil, salt, and pepper and continue to cook until the onion is very soft, about 5 more minutes.

When the potatoes are tender, mash them in their water, and combine them with the onion mixture, along with the diced chilies, corn, and 1 cup of the soymilk. Stir to blend. Add more soymilk if the soup is too thick. Heat gently until hot and steamy.

For a milder soup, use only half of the chilies. Freeze the rest for later use.

Per serving: 179 calories; 4 grams protein; 40 grams carbohydrate; 0.5 gram fat; 305 mg sodium; 0 mg cholesterol

GOLDEN MUSHROOM SOUP

Serves 6

This is a rich-tasting soup, delicious with fresh baked bread, baked yams, and a green salad.

2½ cups water or vegetable stock
2 tablespoons soy sauce
2 medium onions, chopped
1 pound mushrooms, sliced
1½ teaspoons dill weed
1 tablespoon paprika
1 teaspoon caraway seeds (optional)
⅛ teaspoon pepper
1½ cups soymilk
2 teaspoons lemon juice
2–3 tablespoons red wine (optional)

Heat ½ cup of the water or vegetable stock in a large pot. Add 1 tablespoon of the soy sauce along with the chopped onions and cook until the onions are soft and translucent, about 5 minutes.

Add the sliced mushrooms, dill, paprika, caraway seeds, and pepper, and cook another 5 minutes, stirring frequently. Add remaining water or stock and soy sauce, then cover and simmer for 15 minutes.

Place 1 cup of the soup into a blender, along with the soymilk, and blend until smooth. Return it to the pot and heat, without boiling, until steamy. Stir in the lemon juice and red wine just before serving.

Per serving: 81 calories; 3 grams protein; 14 grams carbohydrate; 0.5 grams fat; 229 mg sodium; 0 mg cholesterol

MINESTRONE

Serves 8

This tomato-vegetable soup is delicious as is, or enhance it with additional vegetables of your choosing. Add fresh-baked bread or muffins and a salad for a satisfying meal.

1 small onion, chopped
2 garlic cloves, minced
1 carrot, cut into chunks
2 stalks celery, sliced including tops
2 cups finely shredded green cabbage
2 medium potatoes, scrubbed and cut into chunks
2 tablespoons chopped parsley
4 cups tomato juice
4 cups water or vegetable stock
2 teaspoons dried basil
2 teaspoons mixed Italian herbs
⅛ teaspoon cayenne
1 medium zucchini, diced
½ cup pasta shells
1 15-ounce can cannelini beans, drained
1–2 cups finely chopped kale, collard greens, or spinach

Place the onion, garlic, carrot, celery, cabbage, potatoes, and parsley in a large pot with the tomato juice, water or stock, basil, Italian herbs, and cayenne. Bring to a simmer, then cover and cook 20 minutes.

Add the zucchini, pasta, cannelini beans, and chopped greens, then cover and simmer an additional 20 minutes, until the pasta is tender. Add extra tomato juice or water if the soup becomes too thick.

Per serving; 158 calories; 6 grams protein; 33 grams carbohydrate; 0 grams fat; 347 mg sodium; 0 mg cholesterol

BORSCHT

Serves 8

2 beets, peeled and diced
1 medium onion, chopped
2 medium potatoes, cut in ½-inch cubes
2 carrots, sliced or diced
3 celery stalks, sliced
2 cups finely shredded green cabbage
2 15-ounce cans Swanson's Vegetable Broth
1 15-ounce can crushed tomatoes
4 cups water
1 teaspoon dill weed
½ teaspoon caraway seeds
¼ teaspoon black pepper
1 cup sauerkraut

Combine everything except the sauerkraut in a large pot and bring to a simmer. Cover and cook until the beets and potatoes are tender, about 20 minutes.

Stir in the sauerkraut and serve.

Per serving: 97 calories; 2 grams protein; 22 grams carbohydrate; 0 grams fat; 366 mg sodium; 0 mg cholesterol

GREEN VELVET SOUP

Serves 10

This beautiful green soup is a delicious way to eat your vegetables.

1 onion, chopped
2 stalks celery, sliced
2 potatoes, scrubbed and diced
¾ cup split peas, rinsed
2 bay leaves
6 cups water or vegetable stock
2 medium zucchini, diced
1 medium stalk broccoli, chopped
1 bunch fresh spinach, washed and chopped
½ teaspoon basil
1 teaspoon black pepper
1 teaspoon salt

Place onion, celery, potatoes, split peas, and bay leaves in a large pot with water or stock and bring to a boil. Lower heat, cover, and simmer 1 hour. Remove bay leaves. Add zucchini, broccoli, spinach, basil, and black pepper, and simmer 20 minutes. Transfer to a blender in several small batches and blend until completely smooth, holding the lid on tightly.

Return to pot and heat until steamy. Add salt to taste.

Per serving: 120 calories; 5 grams protein; 24 grams carbohydrate; 0 grams fat; 238 mg sodium; 0 mg cholesterol

CURRIED SWEET POTATO SOUP

Serves 8

Yams take on a whole different character in this spicy, slightly sweet soup.

2 medium yams
½ cup water
1 onion chopped
½ teaspoon each: mustard seeds, turmeric, ginger, and cumin
¼ teaspoon cinnamon
⅛ teaspoon cayenne
¾ teaspoon salt
2 cups water
1 tablespoon lemon juice
2 cups soymilk

Peel the yams and cut them into chunks. Steam over boiling water until tender when pierced with a fork, about 40 minutes. Mash (you should have about 2 cups), and set aside.

Braise the onion in ½ cup water until soft, then add the spices and salt. Cook 2 minutes over low heat, stirring constantly, then whisk in 2 cups water, lemon juice, and mashed yams. Simmer 10 minutes. Add soymilk and puree soup in a blender in two to three batches until very smooth. Return to the pan and heat over a medium flame until hot and steamy, about 10 minutes.

Per serving: 82 calories; 2 grams protein; 18 grams carbohydrate; 0.5 gram fat; 227 mg of sodium; 0 mg cholesterol

OKEFENOKEE STEW

Serves 8

This delicious, easy-to-prepare stew was inspired by a balmy southern evening in the Okefenokee Swamp. For traditional southern flavors, what better than okra and black-eyed peas?

½ cup red wine
1 large onion, chopped
1 bell pepper, diced
2 stalks celery, sliced
3–4 cups sliced okra (¾ to 1 pound)
1 15-ounce can chopped tomatoes
2 teaspoons each: chili powder and oregano
1 tablespoon soy sauce
1 can black-eyed peas, with liquid
2 cups frozen corn

Heat wine in a large pot and cook chopped onion until tender, about 5 minutes. Add celery, bell pepper, okra, tomatoes, and seasonings. Lower heat, cover, and simmer

10 minutes. Add black-eyed peas and corn and simmer 5 to 10 minutes longer. Serve over cooked brown rice.

Per serving: 126 calories; 5 grams protein; 24 grams carbohydrate; 0 grams fat; 324 mg of sodium; 0 mg cholesterol

GRAINS

BROWN RICE

Makes 3 cups of cooked rice

Whole grains are the foundation of a healthful diet, and brown rice is a great grain to start with. It has a crunchy, nutlike flavor and supplies more vitamins, minerals, protein, and fiber than white rice. Cooking it in extra water guarantees perfect rice every time.

1 cup short grain brown rice
4 cups water
½ teaspoon salt

Rinse and drain rice. In a saucepan bring the water to a boil, then add the rice and salt. When the water begins to

161

boil again, lower the heat slightly, then cover and simmer about 40 minutes, until the rice is soft but still retains a hint of crunchiness.

Pour off any excess liquid (this can be saved and used as a broth for soups and stews if desired).

Per ½ cup: 115 calories; 2 grams protein; 25 grams carbohydrate; 0.5 gram fat; 176 mg sodium; 0 mg cholesterol

WILD BASMATI PILAF

Serves 6

Wild rice and basmati rice add wonderful flavor and texture to this fat-free pilaf. Look for both in natural food stores and in some supermarkets. Serve this pilaf with everyday meals or as an elegant side dish.

¼ cup wild rice
1 15-ounce can Swanson's Vegetable Broth
¾ cup brown basmati rice
1 onion, finely chopped
3 garlic cloves, minced
2 cups thinly sliced mushrooms
2 stalks celery, thinly sliced
½ teaspoon thyme
½ teaspoon marjoram
¼ teaspoon black pepper
¼ teaspoon salt
⅓ cup finely chopped parsley

Rinse the wild rice and place it in a saucepan with the vegetable broth and ½ cup of water. Stir to mix, then cover and simmer for 20 minutes. At the end of this time, add the basmati rice.

Cover and continue cooking until both varieties of rice are tender, about 50 minutes.

Heat ½ cup of water in a large pot or skillet. Add the onion and garlic and cook until all the water has evaporated and browned bits of onion begin to stick to the pan. Add another ¼ cup of water, scrape the pan, and cook until the onions begin to stick again. Repeat this process of adding water and cooking the onions until they are nicely browned. This will take about 15 minutes.

Stir in the mushrooms, celery, and seasonings. Cook, stirring frequently, for 5 minutes, then add the cooked rice and finely chopped parsley. Cook over low heat, turning gently, until the mixture is very hot.

Per serving: 124 calories; 3 grams protein; 27 grams carbohydrate; 0.5 gram fat; 284 mg sodium; 0 mg cholesterol

SIMPLE SPANISH RICE

Makes 2½ cups

Here is a simple Spanish rice recipe that's baked in the oven. Spike is a seasoning salt available in most markets in the spice section.

1 cup brown basmati rice (or other long grain brown rice)
2¼ cups boiling water or vegetable stock
2½ tablespoons soy sauce
1 teaspoon chili powder
½ teaspoon Spike seasoning or ¼ teaspoon salt
¼ teaspoon cumin
¼ teaspoon garlic powder

Preheat the oven to 350°. Spread the rice in a 9 × 9–inch baking dish. Combine the boiling water or stock with the remaining ingredients and pour over the rice. Cover tightly with foil and bake in preheated oven until the rice is tender and all the water is absorbed, about 1 hour.

Per ½ cup: 105 calories; 3 grams protein; 23 grams carbohydrate; 0 grams fat; 341 mg sodium; 0 mg cholesterol

BULGUR

Makes about 3 cups

Bulgur is cracked wheat that has been toasted to give it a delicious, nutty flavor. It cooks in about 15 minutes and is an excellent accompaniment to a wide variety of foods, from chili to roasted vegetables. If you like, season it with chili powder, curry powder, or soy sauce. Or you can add it to soups, stews, and chili for extra flavor and texture, serve it as a hot breakfast cereal, or mix it with chopped fresh vegetables and fat-free dressing for a delicious salad.

2 cups water
1 cup bulgur
½ teaspoon salt

Bring water to a boil in a saucepan, then stir in bulgur and salt. Reduce heat to a simmer, then cover and cook until bulgur is tender, about 15 minutes.

Per ½ cup: 113 calories; 4 grams protein; 24 grams carbohydrate; 0 grams fat; 181 mg sodium; 0 mg cholesterol

Alternate method: Stir bulgur and salt together in a mixing bowl. Stir in boiling water. Cover and let stand until the liquid is absorbed and the bulgur is tender, about 20 minutes.

COUSCOUS

Makes 3 cups

Couscous, which is actually a type of pasta, cooks in minutes and makes a delicious side dish or salad. Many natural food stores sell whole wheat couscous, which has a nuttier flavor and more fiber than regular couscous. They are both prepared in the manner described below.

1½ cups boiling water
½ teaspoon salt
1 cup couscous (whole wheat or regular)

Bring salted water to a boil in a small pan. Stir in the couscous, then remove the pan from heat and cover it. Let stand 10 to 15 minutes, then fluff with a fork and serve.

Per ½ cup: 91 calories; 3 grams protein; 20 grams carbohydrate; 0 grams fat; 181 mg sodium; 0 mg cholesterol

POLENTA

Makes 2 cups

Polenta, or coarsely ground cornmeal, has long been a staple grain in northern Italy. It cooks easily and is delicious with marinara or any other spicy sauce. Pour the cooked polenta onto a platter and top it with sauce, or spoon it into a flat dish and chill it, then slice and grill it as described below.

½ cup polenta
1 15-ounce can Swanson's Vegetable Broth
½ cup water

Combine the polenta, vegetable broth, and ½ cup of water in a saucepan. Bring to a simmer and cook, stirring frequently, until very thick, 15 to 20 minutes. Pour into a 9 × 9-inch baking dish and chill completely. Cut into squares and grill or sauté in an oil-sprayed nonstick skillet.

Per ½ cup: 62 calories; 1 gram protein; 14 grams carbohydrate; 0 grams fat; 437 mg sodium; 0 mg cholesterol

QUINOA

Makes 3 cups

Quinoa, pronounced "keen-wah," is a light, fluffy grain that makes a delicious breakfast cereal, side dish, or salad. It cooks in just fifteen minutes and is a rich source of many nutrients, including protein, B vitamins, and iron. Look for quinoa in your favorite natural food store.

    1 cup quinoa
    2 cups water

Rinse the quinoa thoroughly by placing it in a bowl and adding about 3 cups of cold water. Swish it around until the water is cloudy, then pour it into a strainer. Put it back into the bowl and rinse it again. Repeat the process 3 or 4 times until the water remains clear.

Place the rinsed quinoa into a pan and add 2 cups of cold water.

Bring to a boil, then reduce the heat to a simmer. Cover and cook until all the water is absorbed, about 15 minutes.

Per ½ cup: 118 calories; 5 grams protein; 20 grams carbohydrate; 2 grams fat; 3 mg sodium; 0 mg cholesterol

FAVORITE BREAD STUFFING

Serves 8

You'll love this traditional stuffing now that it's fat-free!

½ cup water
1 onion, chopped
3 cups sliced mushrooms (about ½ pound)
2 celery stalks, thinly sliced
4 cups cubed bread
⅓ cup finely chopped parsley
½ teaspoon thyme
½ teaspoon marjoram
½ teaspoon sage
⅛ teaspoon black pepper
½ teaspoon salt
1 cup very hot water or vegetable stock

Heat ½ cup water in a large pot or skillet. Add the onion and cook 5 minutes.

Add the sliced mushrooms and celery and cook over medium heat, stirring occasionally, until the mushrooms begin to brown, about 5 minutes.

Preheat the oven to 350°. Stir in the bread, parsley,

thyme, marjoram, sage, black pepper, and salt. Lower the heat and continue cooking for 3 minutes, then stir in the water or stock, a little at a time, until the dressing obtains the desired moistness. Spread in an oil-sprayed baking dish, cover, and bake for 20 minutes. Remove the cover and bake 10 minutes longer.

Per serving: 91 calories; 3 grams protein; 14 grams carbohydrate; 2 grams fat; 297 mg sodium; 0 mg cholesterol

VEGETABLE DISHES

BRAISED CABBAGE

Serves 4

Adrien Avis provided this simple and delicious vegetable dish.

½ cup water
2 cups cabbage, coarsely chopped
½ teaspoon caraway seeds (optional)
salt and fresh ground black pepper

Bring the water to a boil in a skillet or saucepan and add the cabbage and caraway seeds. Cover and cook until just tender, about 5 minutes. Sprinkle with salt and black pepper and serve.

Per serving: 16 calories; 0.5 gram protein; 4 grams carbohydrate; 0 grams fat; 13 mg sodium; 0 mg cholesterol

STEAMED KALE

Serves 2 to 4

If you've never tried kale, you're in for a treat. Besides being delicious, kale is an excellent source of calcium and beta carotene. This is the simplest method I know of preparing it.

1 bunch of kale (6 to 8 cups chopped)
½ cup water
2 teaspoons soy sauce
2–3 garlic cloves, minced

Wash the kale, remove the stems, and chop the leaves into ½-inch wide strips.

Heat the water and soy sauce in a large pot or skillet. Add the garlic and cook it 1 to 2 minutes. Add the chopped kale and toss to mix. Cover and cook over medium heat until tender, about 5 minutes.

Per 1 cup: 61 calories; 3 grams protein; 11 grams carbohydrate; 0 grams fat; 101 mg sodium; 0 mg cholesterol

POTATOES WITH KALE AND CORN

Serves 6

4 russet potatoes
1 bunch kale
½ cup water
2–3 teaspoons soy sauce
3 garlic cloves, minced
1 15-ounce can corn, including liquid

Scrub the potatoes, cut them into cubes, and steam them over boiling water until tender, about 15 minutes.

Wash the kale and remove the stems. Chop the leaves into small pieces. Heat ½ cup of water and 2 teaspoons of soy sauce in a large pan, then add the garlic. Cook 1 minute, then add the chopped kale. Toss to mix, then cover and cook over medium heat until the kale is tender, about 5 minutes. Add the corn with its liquid, and the cooked potatoes. Stir to mix.

Per serving: 238 calories; 5 grams protein; 52 grams carbohydrate; 0.5 gram fat; 120 mg sodium; 0 mg cholesterol

MASHED POTATOES AND GRAVY

Serves 8

This is a simple, fat-free version of a traditional favorite.

4 large russet potatoes, peeled and diced
2 cups water
½ teaspoon salt
½–1 cup soymilk or rice milk
½ cup water
1 tablespoon soy sauce
½ cup finely chopped onion
1 cup sliced mushrooms
2 tablespoons flour
½ cup water
¼ teaspoon garlic powder or granules
¼ teaspoon poultry seasoning
⅛ teaspoon black pepper

Simmer the potatoes in 2 cups of salted water until tender, about 10 minutes. Drain and reserve the liquid. Mash the potatoes, adding enough milk to achieve desired consistency. Add salt to taste. Cover and set aside.

Heat ½ cup of water and soy sauce in a large skillet, then

add onions and mushrooms. Cover and cook over medium heat 5 minutes, stirring occasionally.

Mix the flour with ½ cup water and add it to the onions along with the seasonings and reserved potato water. Stir over medium heat until thickened.

Per serving: 134 calories; 3 grams protein; 30 grams carbohydrate; 0 grams fat; 228 mg sodium; 0 mg cholesterol

BAKED YAMS

Serves 4

Yams are a great year-round food. Delicious and easy to prepare, they are loaded with beneficial nutrients like beta-carotene, potassium, and fiber. Serve them with meals and keep them on hand for snacks. Eat them plain, or topped with a bit of maple syrup or applesauce.

4 medium yams (jewel and garnet yams are especially flavorful)

Microwave oven: Scrub the yams but do not peel them. Prick them in several places with a fork or sharp knife. Place two yams in the microwave and cook on high for 6 minutes. Turn the yams over and cook an additional 6 minutes. The yams are done when they are easily pierced with a fork.

Conventional oven: Scrub and prick the yams as above. Line the bottom of a baking dish with foil or baking parchment, then place the yams in the dish and bake at 400° for 1 hour, or until tender when pierced with a fork.

Per yam: 158 calories; 1.5 grams protein; 38 grams carbohydrate; 0 grams fat; 12 mg sodium; 0 mg cholesterol

HEARTIER DISHES

SUPER SPAGHETTI

Serves 8

Spaghetti doesn't get any easier, unless you buy it ready-made! This simple, tasty sauce goes well with manicotti, lasagne, or any other pasta dish.

8 ounces spaghetti
½ cup red or white wine or water
1 onion, chopped
4 garlic cloves, crushed
2 cups sliced mushrooms
1 28-ounce can crushed tomatoes
1 tablespoon mixed Italian herbs
1 teaspoon sugar

½ teaspoon fennel seeds (optional)
¼ teaspoon black pepper

Cook the spaghetti until it is tender. Drain and rinse. Set aside.

While the spaghetti cooks, heat the wine or water in a large pot, add the onion, garlic, and mushrooms. Cook until the onion is soft and translucent, about 5 minutes. Add the remaining ingredients and simmer for 15 to 20 minutes.

Spread the cooked spaghetti on a platter and top with the sauce.

Per ½ cup: 103 calories; 4 grams protein; 18 grams carbohydrate; 0 grams fat; 255 mg sodium; 0 mg cholesterol

POLENTA WITH
HEARTY BARBECUE SAUCE

Serves 8

Textured vegetable protein (TVP) provides a meaty flavor with none of meat's fat or cholesterol. You'll find it in natural food stores.

3 teaspoons soy sauce
1 large onion, finely chopped
1 bell pepper, finely diced
3 large garlic cloves, minced
1 15-ounce can tomato sauce
1 cup textured vegetable protein (TVP)
1 tablespoon sugar
1 teaspoon chili powder
2 tablespoons cider vinegar
1 tablespoon peanut butter
1 teaspoon stoneground or Dijon-style mustard
4 cups cooked polenta (1 cup uncooked) (page 169)

Heat ½ cup of water and 1 teaspoon of soy sauce in a large pot.

Add the onion, bell pepper, and garlic and cook until the onion is soft, about 5 minutes. Add the tomato sauce, TVP, sugar, chili powder, vinegar, peanut butter, mustard, and 1 cup of water. Cook over medium heat, stirring frequently, until the TVP is softened and the sauce is thick, about 10 minutes.

Spread the cooked polenta onto a serving bowl or platter, then top with sauce and serve.

Per serving: 146 calories; 8 grams protein; 24 grams carbohydrate; 1 gram fat; 215 mg of sodium; 0 mg cholesterol

SWEET AND SOUR
VEGETABLE STIR-FRY

Serves 4

⅓ cup ketchup
⅓ cup vinegar
⅓ cup brown sugar
1 tablespoon soy sauce
1 tablespoon cornstarch
¼ teaspoon dried red pepper flakes
½ cup water
½ cup water or vegetable stock
1 cup thinly sliced onion
2 garlic cloves, crushed
2 cups sliced mushrooms
1 red bell pepper, thinly sliced
1 medium zucchini, thinly sliced
2 cups snow peas
3–4 cups cooked brown rice (page 161)

Combine the ketchup, vinegar, brown sugar, soy sauce, cornstarch, pepper flakes, and ½ cup of water in a small bowl. Stir to mix, then set aside. In a large skillet or wok, heat ½ cup of water or vegetable stock and add the onion and garlic. Cook until the onion begins to soften, about 3 minutes. Add the mushrooms and cook 3 minutes, then add the bell pepper and zucchini. Continue cooking over medium-high heat, stirring continuously, until the vegetables are just becoming tender, about 3 minutes. Add the snow peas and reserved sauce mixture and cook, stirring constantly, until the sauce is clear and thickened, about 2 minutes.

Serve over cooked rice.

Per serving: 352 calories; 8 grams protein; 73 grams carbohydrate; 3 grams fat; 320 mg sodium; 0 mg cholesterol

FAST LANE CHOW MEIN

Serves 4 to 6

This chow mein takes just minutes to prepare and is a delicious way to load up on nutritious vegetables.

- ½ cup water
- 2 teaspoons soy sauce
- 1 onion, chopped
- 4–5 garlic cloves, minced
- 1 cup thinly sliced green cabbage
- 1 stalk celery, thinly sliced
- 1 pound sliced mushrooms, fresh or frozen
- 3 cups chopped kale
- 2 packages vegetarian ramen soup (I use Westbrae Miso Ramen)
- 1 cup water

Heat ½ cup of water and 1 teaspoon of soy sauce in a large pot. Add the chopped onion and garlic. Cook until the onion is soft, about 5 minutes. Add the cabbage, celery, mushrooms, and kale.

Crush the ramen noodles (I use the handle of my knife to do this) and add them to the vegetable mixture along with one of the ramen seasoning packets and 1 cup of water. Stir to mix, then cover and cook over medium heat, stirring occasionally, until the vegetables and noodles are tender, about 5 minutes. Check the seasonings and add a bit more of the ramen seasoning mix if needed.

Per serving: 163 calories; 6 grams protein; 32 grams carbohydrate; 1 gram fat; 594 mg sodium; 0 mg cholesterol

THAI VEGETABLES WITH RICE

Serves 8

In this dish, colorful vegetables are simmered in a flavorful sauce and served over aromatic basmati or jasmine rice. For a mild dish, use the smaller amount of dried red pepper flakes. Increase the amount for a more fiery version.

½ cup water
2 tablespoons soy sauce
1 onion, thinly sliced
4 garlic cloves, minced
1 pound yams, peeled and cut into wedges
1 15-ounce can crushed tomatoes
2 teaspoons ground coriander
1 teaspoon ground cumin
½–1 teaspoon red pepper flakes
½ teaspoon turmeric
½ teaspoon ginger powder (or 2 teaspoons fresh ginger, grated)
½ cup water
1 15-ounce can garbanzo beans, including liquid
½ pound fresh green beans, trimmed and cut into 1-inch pieces (about 2 cups)

1 red bell pepper, cut into thin strips
2 teaspoons grated lemon rind (lemon zest)
1 tablespoon lemon juice
1 tablespoon soy sauce
6 cups cooked basmati or jasmine rice (2 cups raw)

Heat ½ cup of water and 1 tablespoon of soy sauce in a large pot. Add the chopped onion and garlic. Cook until the onion is very soft, about 5 minutes. Add the yam, crushed tomatoes, coriander, cumin, red pepper flakes, turmeric, ginger, and ½ cup of water. Cover and simmer 5 minutes.

Add the garbanzo beans with their liquid and the green beans, then cover and simmer 5 more minutes. Add the bell pepper and lemon rind and simmer until the yams and green beans are tender when pierced with a sharp knife, about 10 minutes. Stir in the lemon juice and remaining soy sauce. Serve over cooked rice.

Per serving: 286 calories; 7 grams protein; 62 grams carbohydrate; 1 gram fat; 166 mg sodium; 0 mg cholesterol

CHILI MAC

Serves 8

Kids love this easy-to-make meal.

- 8 ounces pasta spirals
- ½ cup water
- 1 onion, chopped
- 1 small bell pepper, diced
- 3 garlic cloves, minced
- ¾ cup textured vegetable protein
- 1 28-ounce can crushed tomatoes
- 1 15-ounce can kidney beans, including liquid
- 1 15-ounce can corn, including liquid
- ¾ cup water
- 2 teaspoons ground cumin
- 4–6 teaspoons chili powder
- ½ teaspoon salt

Cook the pasta in boiling water until it is tender. Drain and rinse it, then set it aside.

Heat ½ cup of water in a large pot then add the onion, bell pepper, and garlic. Cook until the onion is soft, about 5 minutes. Add the textured vegetable protein, crushed tomatoes, kidney beans, corn, and an additional ¾ cup of water. Stir in the cumin, chili powder, and salt, then simmer over medium heat, stirring occasionally, for 20 minutes. Stir in the cooked pasta and check the seasonings. Add more chili powder if a spicier dish is desired.

Per serving: 170 calories; 9 grams protein; 32 grams carbohydrate; 1 gram fat; 434 mg sodium; 0 mg cholesterol

FAT-FREE HOT DOGS AND BURGERS

Now you and your family can enjoy delicious vegetarian hot dogs, chili dogs, and burgers without any fat or cholesterol. Look for Smart Dogs, Yve's Veggie Wieners, or Garden Dogs in your supermarket or natural food store. Delicious vegetarian burger patties are available in many supermarkets and natural food stores.

Brands vary in flavor and texture, so do some exploring to find your favorites. Some of my favorites are the Garden Vegan by Wholesome and Hearty Foods, the Boca Burger by the Boca Burger Company, and the Natural Touch Burger by Morningstar Farms. Be sure to select varieties that are low in fat (3 grams or less per patty). You may also want to try using vegetarian burgers in place of meat in sauces, stuffings, and tacos (see the recipe for tacos on page 194).

CHILI BURGER DELUXE

Makes enough for 2 burgers

Use one of the delicious fat-free, meatless patties listed above for this great-tasting burger.

2 vegetarian burger patties
1 burger bun, preferably whole grain
1 15-ounce can vegetarian chili
2 tomato slices
barbecue sauce or salsa

For each burger you are making, heat one patty in a toaster oven, microwave, or in a skillet, lightly toast one half of a burger bun, and heat about 1 cup of chili on the stove or in a microwave.

To assemble, top the bun half with a burger patty, a tomato slice, and the heated chili. Garnish with barbecue sauce or salsa.

Per burger: 352 calories; 24 grams protein; 60 grams carbohydrate; 1 gram fat; 566 mg sodium; 0 mg cholesterol

TACOS

Makes 12 tacos

Boca Burgers are fat-free, meatless patties, available in natural food stores and in a growing number of supermarkets. If you are unable to locate Boca Burgers, substitute any other fat-free vegetarian burger.

½ cup water
½ onion, chopped
3 garlic cloves, crushed
1 small bell pepper, finely diced
½ cup tomato sauce
4 Boca Burgers, thawed and crumbled
2 teaspoons chili powder
½ teaspoon cumin
½ teaspoon oregano

12 corn tortillas
1 cup shredded leaf lettuce
1 medium tomato, diced
4 green onions, sliced
½ cup salsa or taco sauce

Heat ½ cup of water in a large skillet and cook onion, garlic, and pepper until the onion is soft, about 5 minutes. Stir in the tomato sauce, crumbled Boca Burgers, chili powder, cumin, and oregano. Cook over low heat until the mixture is fairly dry, about 3 minutes.

To assemble the tacos, place a small amount of filling on a tortilla and place it in a heavy, ungreased skillet. Heat until the tortilla is soft and warm, then fold in half and cook on each side for 60 seconds or longer for a crisper taco. Garnish with remaining ingredients.

Per taco: 114 calories; 7 grams protein; 19 grams carbohydrate; 1 gram fat; 86 mg sodium; 0 mg cholesterol

BROCCOLI BURRITOS

Makes 6 burritos

This delicious dish turns broccoli into a meal. Roasted red peppers are available in jars in most markets, and tahini is sold in natural food stores, ethnic markets, and many supermarkets.

    1 bunch of broccoli
    1 15-ounce can garbanzo beans
    ½ cup roasted red peppers
    2 tablespoons tahini (sesame seed butter)
    3 tablespoons lemon juice
    6 flour tortillas
    6 tablespoons salsa (more or less to taste)

Cut or break the broccoli into flowerets. Peel the stalks and cut them into ½-inch thick rounds. You should have about 2 cups. Steam over boiling water until just tender, about 5 minutes.

Drain the garbanzo beans and place them in a food processor with the peppers, tahini, and lemon juice. Process until very smooth.

Spread about ¼ cup of the garbanzo mixture on a tortilla and place it face up in a large heated skillet. Heat it until the tortilla is warm and soft, about 2 minutes. Spread a line of broccoli down the center of the tortilla and sprinkle with salsa.

Fold the bottom of the tortilla up, then starting on one side, roll the tortilla around the broccoli. Repeat with the remaining tortillas.

Per burrito: 244 calories; 9 grams protein; 39 grams carbohydrate; 5 grams fat; 130 mg sodium; 0 mg cholesterol

Tip: You can reduce the fat content by using fat-free flour tortillas, which are sold in many supermarkets. Or try using chapatis (Indian flatbreads), which are made with whole wheat flour and are usually fat-free. Look for them in your favorite natural food store.

QUICK BLACK BEAN CHILI

Serves 6

There are so many ways to enjoy delicious Black Bean Chili: serve it with brown rice and a green salad, use it as a filling for burritos, or try it over potatoes, as suggested in the next recipe. For a hotter chili, add some cayenne or fresh chopped jalapenos.

½ cup water
1 tablespoon soy sauce
2 onions, chopped
4 garlic cloves, crushed
2 15-ounce cans black beans
1 15-ounce can crushed tomatoes
1 4-ounce can diced chilies
2 teaspoons oregano
½ teaspoon cumin
¼ teaspoon salt
fresh cilantro, chopped

Heat the water and soy sauce in a large pan, then add the onion and garlic and cook until the onion is soft, about 5 minutes.

Add the black beans with their liquid, the tomatoes, chilies, oregano, cumin, and salt. Cover and simmer 25 minutes. Top with chopped fresh cilantro and serve.

Per serving: 138 calories; 7 grams protein; 26 grams carbohydrate; 0 grams fat; 199 mg sodium; 0 mg cholesterol

POTATOES WITH
BLACK BEAN CHILI

Makes 4 potatoes

4 red or yellow potatoes
1 cup black bean chili, homemade or commercial
½ cup salsa
2 green onions, sliced
¼ cup chopped cilantro

Scrub the potatoes and steam them until they are just tender when pierced with a fork, about 35 minutes. Cut a slit in the potato and press on the ends to open it. Top with black bean chili, salsa, green onions, and cilantro.

Per serving: 376 calories; 11 grams protein; 82 grams carbohydrate; 0.5 gram fat; 450 mg sodium; 0 mg cholesterol

BAKED FALAFEL SANDWICH

Makes 20 falafel patties

Stuff a piece of whole wheat pita bread with these baked falafel patties for a delicious "Middle Eastern Taco."

1 potato, diced
1 15-ounce can garbanzo beans, drained
1 small onion, finely chopped
½ cup fresh parsley, chopped fine
2 garlic cloves, minced
2 tablespoons tahini (sesame seed butter)
2 tablespoons soy sauce
½ teaspoon turmeric
½ teaspoon coriander
1 teaspoon cumin
⅛ teaspoon cayenne
Pita bread
Garnishes: lettuce, cucumber, tomato, green onions
salsa (optional)

Steam the diced potato over boiling water until soft. Place the drained garbanzo beans in a bowl, add the cooked potato, and mash. Add the onion, parsley, garlic, tahini, soy sauce, turmeric, coriander, cumin, and cayenne. Mix well.

Preheat the oven to 350°. Form the garbanzo mixture into small (walnut-sized) patties and place on a greased baking sheet. Bake for 15 minutes, then turn with a spatula and bake an additional 15 minutes.

Warm the pita bread in foil in the oven, or by steaming it, or heating it briefly in a microwave oven. Stuff two or three falafel patties into the warm pita bread and garnish with lettuce, cucumber, tomato, and green onions. Top with salsa if desired.

Per falafel: 214 calories; 8 grams protein; 39 grams carbohydrate; 2 grams fat; 260 mg of sodium; 0 mg carbohydrate

Tip: Fantastic Foods makes an excellent instant falafel mix: just add water and let it stand a few minutes, then form into patties and bake as above.

PASTA WITH CREAMY PESTO

Serves 8

This is pesto that you can enjoy to your heart's content. It's made with silken tofu and contains no added oil.

1 12-ounce package pasta
¼–½ teaspoon salt
1 10.5-ounce package firm silken tofu
2 cups fresh basil leaves
1 garlic clove
½ teaspoon salt

Cook the pasta in boiling water until it is tender, then drain and rinse it. Sprinkle with ¼ teaspoon of salt and toss to mix.

While the pasta is cooking, combine the remaining ingredients in a food processor. Blend until completely smooth, stopping the processor occasionally and scraping down the sides with a rubber spatula.

Add the pesto to the pasta and toss gently to mix. Serve at once.

Per serving: 158 calories; 8 grams protein; 28 grams carbohydrate; 1 gram fat; 204 to 270 mg of sodium; 0 mg cholesterol

CURRIED MUSHROOMS AND CHICKPEAS

Serves 8

This spicy Indian dish is delicious with rice or couscous. You can make it hotter or milder by adjusting the amount of cayenne.

½ cup water
2 large onions, chopped
1½ tablespoons cumin seeds
1½ pounds mushrooms, sliced
1 28-ounce can crushed tomatoes
1 15-ounce can garbanzo beans, drained
1 teaspoon turmeric
1 teaspoon coriander
½ teaspoon cayenne
½ teaspoon ginger

Heat the water in a large pot, then add the onions and cook until they are soft. Add the cumin seeds and mushrooms, and continue cooking over medium heat until the mushrooms are light brown.

Add the tomatoes, garbanzo beans, and spices. Cook 30 minutes or longer, until the mushrooms are tender and the flavors are well blended.

Per serving: 116 calories; 5 grams protein; 20 grams carbohydrate; 1 gram fat; 286 mg sodium; 0 mg cholesterol

DESSERTS

GINGER PEACHY BREAD PUDDING

Serves 9

1 28-ounce can sliced peaches
1 tablespoon cornstarch
6 cups cubed whole grain bread (about 8 slices)
1¾ cups soymilk or rice milk
⅓ cup packed brown sugar
¾ cup golden raisins
½ teaspoon ginger
½ teaspoon cinnamon
¼ teaspoon nutmeg
¼ teaspoon salt
1 teaspoon vanilla

¼ cup finely chopped crystallized ginger (optional)
2 tablespoons brown sugar

Drain the liquid from the peaches into a large mixing bowl and add the cornstarch. Stir to dissolve any lumps, then add the bread cubes, soy or rice milk, brown sugar, raisins, ginger, cinnamon, nutmeg, salt, and vanilla. Mix well. Stir in the crystallized ginger if desired. Chop the peaches and stir them to the mixture. Spread in a 9 × 9–inch baking dish, then sprinkle the top with brown sugar and let stand 5 minutes while the oven preheats to 350°. Bake for 35 minutes. Serve warm or cooled.

Per serving: 252 calories; 3 grams protein; 57 grams carbohydrate; 1 gram fat; 185 mg sodium; 0 mg cholesterol

PEACH COBBLER

Serves 9

Fresh peaches are the essence of summertime, and this is such a delicious way to eat them. If you get a yearning for this cobbler in the middle of winter, use frozen peaches instead.

½ cup sugar
2 tablespoons cornstarch
1 cup water
5 cups sliced peaches, fresh or frozen
½ teaspoon cinnamon
1¼ cups whole wheat pastry flour
2 tablespoons sugar
1½ teaspoons baking powder
¼ teaspoon salt
½ cup soymilk or rice milk

Mix the sugar and cornstarch in a saucepan, then stir in the water and peaches. Bring to a boil and cook over medium-high heat, stirring constantly, until the sauce is clear and thick.

Pour into a 9 × 9–inch baking dish, and sprinkle with

the cinnamon. Preheat the oven to 375°. Mix the flour, 2 tablespoons of sugar, baking powder, and salt in a large bowl, then stir in the soymilk or rice milk to form a batter. Drop by spoonfuls onto the hot peach mixture. Bake until golden brown, about 30 minutes.

Per serving: 178 calories; 3 grams protein; 40 grams carbohydrate; 0 grams fat; 73 mg sodium; 0 mg cholesterol

CHOCOLATE TORTE

Serves 12

This torte is pure chocolate decadence. It tastes unbelievably rich, yet is surprisingly low in fat.

In a medium saucepan, stir together the following.

1 cup couscous
1 cup sugar
¼ cup cocoa
¼ teaspoon salt
2½ cups water

Bring to a simmer and cook over medium heat until thickened, about 7 minutes. Spread in the bottom of an ungreased 9-inch springform pan.

In another medium saucepan, combine the following.

½ cup sugar
5 tablespoons cornstarch
3 tablespoons cocoa
2 cups soymilk or rice milk

Whisk smooth, then cook over medium heat, stirring constantly until very thick (like thick pudding). Spread evenly over the top of the couscous mixture.

Refrigerate until set, at least 2 hours, before serving.

Per serving (¹⁄₁₂th of pie): 161 calories; 3 grams protein; 36 grams carbohydrate; 1 gram fat; 106 mg sodium; 0 mg cholesterol

GINGERBREAD

Makes 9 two-inch pieces

1⅔ cups unbleached or whole wheat pastry flour
1 teaspoon baking soda
½ teaspoon salt
1½ teaspoons cinnamon
1½ teaspoons ginger
1 10.5-ounce package firm silken tofu
½ cup sugar
¼ cup molasses
½ cup raisins

Preheat oven to 375°. Mix the flour, baking soda, salt, cinnamon, and ginger in a large bowl.

Place the tofu, sugar, and molasses into a food processor and blend until completely smooth. Add to the flour mixture and stir until just blended. Mix in the raisins. Spread in an oil-sprayed 8-inch square baking dish, and bake in the preheated oven for 25 minutes. Use a spatula to loosen the gingerbread from the pan, then turn it out onto a rack to cool.

Per muffin: 198 calories; 6 grams protein; 41 grams carbohydrate; 1 gram fat; 224 mg sodium; 0 mg cholesterol

POWER SNACKS

Snacking does not have to mean gaining weight. With practical tips and delicious recipes, including sweets, savories, and beverages, "Power Snacks" lets you say "yes" to your taste buds and still lose weight. Using the incredible negative calorie effect, these snacks contain nutrients that help keep your metabolism burning calories faster for hours.

You have been really good to yourself—eating exactly right, taking advantage of the foods that melt away the pounds. You can almost feel the weight dissolving away. And you have the confidence of knowing that success at permanent weight control is finally within your reach.

Then one night you make a trip to the kitchen and open the cupboard. To your horror, it is filled with chocolate doughnuts, potato chips, and cheesy snacks staring at you menacingly.

You slam the cupboard door closed. "No way," you think to yourself. "I'm not going to destroy all my progress with that junk food." You turn to the refrigerator. And there, sliced bologna, cheese, and frozen pizza lie in wait for their next victim. You suddenly feel success slipping through your fingers like sticky melted ice cream.

What is a late-night snacker to do—yield to temptation, knowing you'll regret it later, or say no and go to bed hungry?

When I was in college, my friends and I often made late-night raids on the local "greasy spoon" café for monster burgers and onion rings, all dripping in grease. And we have all found ourselves opening up a packet of cookies and eating far more than we wish we had, or polishing off a carton of ice cream, or maybe throwing aside any semblance of delicacy and just tucking into a big jar of peanut butter with a spoon. More often than we'd like, we have been taken a bit too far by our taste buds, making our waistlines pay the price.

There is a solution. When you have the right kind of snacks on hand, you can kiss those fatty varieties goodbye, along with the pounds they would bring.

Following are simple recipes that let you treat yourself to all kinds of wonderful snacks, including cakes, pies, and cookies that are delicious, but totally healthy and in sync with your weight-control program. If your taste calls for pizza, dips, or french fries, you'll find easy ways to make them that actually help you lose weight, because they help you burn calories.

Before turning to the recipes, let me share with you a few thoughts about snacking.

First of all, don't be afraid of snacking. The old-fashioned frowning on between-meal snacks has been replaced by the recognition that more frequent meals or snacks can actually be beneficial in some ways, compared to fewer, bigger meals. If, on the other hand, you are intentionally eating small meals to try to lose weight only to need snacks later to deal with the inevitable hunger, go ahead and have more generous meals.

When it comes to snacking, there seems to be a gender difference in the kinds of foods we are drawn to. Women often go for sweets, particularly chocolate, while men tend to seek out fried or salty items, such as potato chips, peanuts, or onion rings, although some men do like sweets and some women prefer savories. It helps to know your tendencies and vulnerabilities, so you can be sure to have healthy varieties of your kind of snack on hand.

If you are drawn to sweets, you'll be glad to know that sugar, used in modest amounts, is not usually the cause of weight problems. It actually has only 4 calories per gram (a gram is about ⅟₂₈ of an ounce). Compare that to the shortening or butter in a typical cookie or cake. Any kind of fat, whether it comes in butter, shortening, margarine, vegetable oil, or any other variety, has 9 calories in every gram, more than twice the calories of sugar.

Where sugar does its dirty work is when it acts as a Trojan horse for fat. Sugar lures you in to cakes, cookies, and pies, but it is the butter or shortening baked into them that will expand your waistline.

Of course, if you eat a heroic amount of sugar day after day, you are likely to gain weight. But the point is that the fats and oils in foods are much more likely to cause weight problems than whatever sugar they may contain.

Nature has always offered us sweet snacks that are as healthy as they are delicious. You can't beat fruits for good nutrition, and even though they are naturally sweet, it is almost impossible to gain weight from them. Open up a big, ripe orange. Or bite into a crisp, fresh apple. Enjoy pears, cherries, pineapples, grapefruit, mangoes—you name it.

If you cut a pineapple or watermelon into chunks and put it in the refrigerator on Sunday, you will be delighted to see it on Monday when you get home from work. Or, for convenience, pick up some fruit cocktail or fruit salad at the grocery store.

Fresh fruit is also a handy thing to have in your desk drawer at work, and dried fruit makes a good snack, whether it is the mixed variety sold in vending machines and gift shops, dried apricots, raisins, or whatever.

Speaking of snacking at work, you might also try the new instant soups, which come in a wide variety of flavors. You'll also find couscous, chili, and bean and rice dishes in the same instant cup package. With some rice cakes or low-fat crackers, you've got a hearty snack or even a complete meal, which you can follow with an apple or orange for dessert.

Some commercial products are worth mentioning. Health food stores and kosher groceries carry Tofutti, which has the taste and texture of the finest ice cream, but is made from soy, so it contains no animal fat, animal protein, or lactose sugar. Look for the "lite" varieties. Health food stores carry many other ice cream and yogurt substitutes made from soy or rice, and they are a big improvement over the cow's milk varieties. Read the labels and choose the lowest-fat brands.

Health food stores also offer new varieties of potato chips and tortilla chips. Typical potato chips are made by dropping potato slices into a fryer. As the hot oil soaks in, their calorie content triples. Several companies now offer potato chips that are baked, not fried, so no oil is added. The same goes for tortilla chips, and as you dip them into salsa or bean dip, you'll never know the difference.

While you're at the health food store, take a look at fat-free cookies. You'll also find many varieties of meatless hot dogs, burgers, and deli slices. Read the labels to pick the lowest-fat brands, but they all beat the socks off the meat varieties. You may want to keep some in your fridge for when you get a serious snacking urge.

Grains work their way into healthy snacks, too. A piece of toast or some noodle soup provides plenty of fat-burning complex carbohydrates. Two important tips: don't put butter or margarine on your toast. Top it with jam or cinnamon instead. And go for whole grains that retain the natural fiber. That means whole grain bread or pasta, brown rice, etc.

We don't usually think of popcorn as health food, but it really is a healthy, natural grain. And when it is air-popped, no fat is added. The danger comes if you drip butter all over the top, and the popcorn simply becomes a vehicle for grease that will fatten you up.

If you have a little extra time on a weekend, you may want to make a snack for the following week—a cake, a pie, or some cookies. This is only for those who know they will be snacking and want a healthier version. Don't bother if you would not normally look for a snack.

Which brings me to another important point: Although the foods here are just about the healthiest snacks around, the idea is not to gorge yourself, shoveling in one great snack after another. Listen to your natural hunger and satiety cues. If you are hungry, that is a time to eat. When you are full, that is a time to do something else. If you are eating simply because you have nothing else to do, all the snacks in the world will not provide the answer you really need. It is time to explore intellectual or physical "nourishment," if you will—the company of friends, good books, films, sports, games, dancing, or whatever. Don't let food take the place of other parts of life.

Many of our between-meal trips to the refrigerator or the store are for beverages rather than food, so let's talk for a minute about these pourable snacks.

The fact is, as far as beverages go, all the body really needs is water, and it is a lot healthier than sodas. Sodas contain either sugar or artificial sweeteners, neither of which your body really needs. NutraSweet, in particular, has been the subject of scientific reports linking it to major health problems. While the toxicologists try to sort out whether it is safe or not, it is worth noting that NutraSweet has been no magic wand for weight problems.

But compared to cola commercials showing teenagers partying on the beach, a glass of water seems pretty dull. Luckily, water has taken on a new sophisticated image, thanks to Evian and Perrier. After all, what cola comes from underground springs in the French Alps? To dress it up, use a squirt of juice from a freshly cut lemon or lime. If you like a thick, sweet shake, try the recipes for fruit smoothies. Kids love them, and you will, too.

If your idea of a healthy beverage is a jolt of caffeine, you will be glad to know that one or two cups of coffee per day does not have serious health consequences, so far as anyone knows, although caffeine can make premenstrual syndrome worse, and if you have more than two cups per day, the caffeine will encourage the loss of calcium from your bones.

Speaking of health concerns, you may have already been paying close attention to the kinds of snacks you eat because of their effect on your cholesterol level, for example. You will be glad to know that all the snacks in this book have no cholesterol at all and tend to be low in fat.

If you have diabetes, you already know that sugary snacks can make it hard to manage your blood sugar, but are necessary when your blood sugar gets too low. You will find that avoiding fatty snacks helps, too, since reducing fat intake helps insulin to work better. Also, snacks made from beans and other legumes often have less effect on blood sugar than other foods. Try white bean dip with fresh ginger and lime, hummus, or bean burritos, or open a can of lentil soup. Whole foods, such as brown rice or whole-grain breads, are better than refined foods, such as white rice or white bread.

If you have high blood pressure, you have probably been shying away from salty snacks, like potato chips and pretzels, and for good reason. Salt can increase your blood pressure. But it is just as important—maybe even more important—to avoid fatty snacks. When people go on very-low-fat diets, especially vegetarian diets, their blood pressure tends to drop. So don't have chicken salad or beef jerky as your appetite-quencher. Have the healthy, low-fat snacks you'll see in this book.

I hope you enjoy these healthy power snacks.

WHITE BEAN DIP WITH FRESH GINGER AND LIME

Serves 6

Minced ginger and fresh lime add a sparkling freshness to this low-fat bean dip. It is delicious with raw vegetables, crackers, or as a spread for crusty whole wheat bread.

3 cups cooked white, navy, or cannellini beans, or 2 cans
    drained rinsed beans
1 tablespoon olive oil (optional)
zest and juice of 1 lime (scrape off the zest before you
    squeeze the juice)
2 or 3 garlic cloves, peeled and minced or pressed through a
    garlic press
1 1-inch piece fresh ginger root, peeled and minced
2 to 3 green onions, minced, including some green tops
dash of hot sauce
¼ teaspoon salt (optional)
3 tablespoons chopped cilantro

In a blender or food processor, process the beans, olive oil, lime zest and juice, garlic, minced ginger, green onion, hot sauce, and salt until smooth. Transfer to a medium bowl and add chopped cilantro. Serve either chilled or at room temperature.

PITA PIZZAS

Makes 6 pizzas

Pita bread makes a terrific crust for quick and easy individual pizzas that you can make in a regular oven or a toaster oven. You can use a commercial pizza sauce or make your own with the following recipe.

1 15-ounce can tomato sauce
1 6-ounce can tomato paste
1 teaspoon each: garlic powder and dried basil
½ teaspoon each: dried oregano and thyme
6 pita breads
2 cups chopped vegetables: green onion, bell pepper,
    mushrooms, and/or olives

Preheat the broiler.

Combine tomato sauce, tomato paste, and seasonings. This will make about twice as much sauce as you need. The extra may be refrigerated or frozen for future use.

To assemble the pizza, turn pita bread upside down so it looks like a saucer. Spread with tomato sauce, then top liberally with chopped vegetables. Place on a cookie sheet and broil about 5 minutes, or until the edges just start to get crisp. Or place individual pizzas in the toaster oven and toast for 3 to 5 minutes.

HUMMUS

Serves 6 to 8

Hummus (pronounced HUMM-us) is a delicious Middle-Eastern chickpea pâté served with crackers, wedges of pita bread, or fresh vegetable slices. It also makes a delicious sandwich spread. It is easy to make in a food processor—just add all the ingredients and process until smooth—or by hand. Store it in an airtight container in the refrigerator for up to a week for quick sandwiches and snacks.

2 cups cooked chickpeas, or 1 15-ounce can
1–2 garlic cloves, minced
¼ cup tahini (sesame seed butter)
2 tablespoons lemon juice
¼ teaspoon salt
1 tablespoon finely chopped fresh parsley
¼ teaspoon each: ground cumin and paprika

Drain the chickpeas, reserving the liquid. Mash the beans, then add the remaining ingredients and mix well. The texture should be creamy and spreadable. If it is too dry, add enough of the reserved bean liquid to achieve the desired consistency. For a fat-free version, replace the tahini with 1 finely grated carrot.

WHOLE WHEAT HERBED
PITA CHIPS

Separate medium-sized whole wheat pita bread into halves. Then cut each half into eighths. Arrange on a cookie sheet, spray lightly with olive oil Pam and dust with garlic salt and mixed crushed dried herbs. Thyme and oregano make an especially good combination.

WHOLE WHEAT CRACKERS
WITH CINNAMON SUGAR

Look for Ak-Mak whole wheat crackers at the grocery store. They are delicious whole wheat, nonfat crackers that are great with dips. They can also satisfy a sweet tooth. Place them on a cookie sheet, spray lightly with Pam or other light oil spray, and dust with cinnamon sugar. Bake at 350 degrees for 5 minutes. This recipe also works with RyKrisp or any similar cracker.

WHOLE WHEAT CRACKERS
WITH HERBS

Follow recipe directions on page 226, but spray with an olive oil spray and lightly dust on your favorite Italian herb blend and a dash of garlic salt.

QUICK BLACK BEAN DIP WITH CORN TORTILLA CHIPS

Fantastic World Foods makes a delicious Black Bean Dip that you'll find at most groceries and at all health food stores. Just follow the directions on the box.

You will also find no-fat, baked tortilla chips at health food stores. Or try this easy way of making your own: Cut a stack of tortillas into eighths and toast the wedges on an ungreased cookie sheet until lightly tanned. Watch carefully—they seem to brown just when you look the other way. For added flavor, give them a quick spray with Pam and sprinkle on an herb and spice blend seasoning.

QUICK VEGETABLE RAMEN

Serves 2

You'll find ramen soups in many different flavors at health foods stores and most supermarkets. The package contains dried noodles that cook in 2 to 3 minutes and a packet of flavorful seasoning broth. By adding your own fresh vegetables, you can dress ramen noodles up a bit, and make a tasty, nutritious snack or meal.

1 package ramen soup
1 cup chopped broccoli
1 green onion, sliced

Follow the package instructions for cooking ramen. Add the broccoli to the boiling water along with the noodles. Stir in the sliced green onion just before serving.

CARROT AND RAISIN SALAD

Serves 6

This is a great snacking salad that will stay fresh and crisp for several days if tightly covered and refrigerated. Grate the carrots and toss with remaining ingredients. Hold the sunflower seeds until just ready to serve, and go lightly, as they do contain fat.

1½ lb carrots
2 teaspoons canola oil
Juice from 1 large lemon
1 tablespoon raspberry vinegar
½ cup raisins
1 tablespoon roasted sunflower seeds (optional)

FOUR BEAN SALAD

Serves 10

This quick salad keeps well in the refrigerator for a tasty snack.

- 1 15-ounce can dark kidney beans, drained
- 1 15-ounce can black-eyed peas, drained
- 1 10-ounce package frozen lima beans, thawed
- 1 15-ounce can S & W Pinquitos or other vegetarian chili beans
- 1 large red bell pepper, diced
- ½ cup finely chopped onion
- 2 cups fresh or frozen corn
- ¼ cup seasoned rice vinegar
- 2 tablespoons apple cider or distilled vinegar
- 1 lemon, juiced
- 2 teaspoons cumin
- 1 teaspoon coriander
- ⅛ teaspoon cayenne

Drain the kidney beans, black-eyed peas, and lima beans and combine in a large bowl. Add the pinquitos or chili beans along with their sauce. Stir in the bell pepper, onion, and corn.

Whisk the remaining ingredients together and pour over the salad. Toss gently to mix. Chill at least 1 hour before serving, if possible.

MOCK TUNA SALAD

Makes 4 sandwiches

This is an amazingly fast way to whip up a healthy sandwich spread, and it has none of the fat, mercury, or other undesirables found in tuna fish.

    1 15-ounce can chickpeas, drained
    1 stalk celery, finely chopped
    1 medium carrot, grated (optional)
    1 green onion, finely chopped
    2 teaspoons Dijon mustard or eggless mayonnaise
    1 tablespoon sweet pickle relish
    ¼ teaspoon salt (optional)

Mash the chickpeas with a fork or potato masher. Leave some chunks. Add the celery, carrot, green onion, mustard, and relish. Add salt to taste.

Serve on whole wheat bread or in pita bread with lettuce and sliced tomatoes.

EGGLESS SALAD SANDWICH

Makes 4 sandwiches

Do you love egg salad, but hate the fat and cholesterol? Here's a better way.

½ pound firm tofu, mashed
1 green onion, finely chopped
2 tablespoons eggless mayonnaise
1 tablespoon pickle relish
1 teaspoon mustard
¼ teaspoon each: ground cumin, turmeric, and garlic
    powder
Pinch of salt

Combine all ingredients and mix thoroughly. Serve on whole wheat bread with lettuce and tomato.

BEAN BURRITO WITH SALSA

Makes 6 burritos

A burrito makes a great snack or a quick meal. By the way, the term "refried beans" is really a misnomer. They arc boilcd, and fat-free and vegetarian varieties are widely available.

    1 15-ounce can refried beans
    1 16-ounce jar salsa
    1 package whole wheat tortillas

Heat the beans in one pan and the salsa in another. Heat a tortilla in a dry, heavy skillet over moderate heat until it is warm and flexible. Remove from pan and spread refried beans in a line down the middle of the tortilla. Fold in the ends, then starting at one side, roll up around the beans. Place on a plate, then spoon heated salsa over the top.

QUICKIE QUESADILLAS

Makes 12 quesadillas, 6 servings

These quesadillas are made with Cheezy Garbanzo Spread. If you make the spread in advance, the quesadillas can be prepared in a jiffy.

1 recipe Cheezy Garbanzo Spread (recipe follows)
12 corn tortillas
3 to 4 green onions, sliced
1 bell pepper, seeded and diced (optional)
2 cups diced tomatoes (optional)
1 cup salsa

Spread 2 to 3 tablespoons of the garbanzo spread on a tortilla and place it, spread side up, in a large heated skillet. As soon as it is warm and soft, fold it in half, then cook it another minute. Remove it from the pan and carefully open it. Sprinkle on some green onions, bell pepper, tomatoes, and salsa. Repeat with the remaining tortillas.

CHEEZY GARBANZO SPREAD

Makes about 2 cups, 8 ¼-cup servings

Try this spread on bread and crackers, or in the quesadilla recipe above. Look for jars of water-packed roasted red peppers near the pickles and olives in your grocery store. Tahini is available in most supermarkets and health food stores.

  1 15-ounce can garbanzo beans
  ½ cup roasted red peppers
  3 tablespoons tahini (sesame seed butter)
  3 tablespoons lemon juice

Drain the garbanzo beans, reserving the liquid, and place them in a food processor or blender with the remaining ingredients. Process until very smooth. If using a blender, you will have to stop it occasionally and push everything down into the blades with a rubber spatula. The mixture should be quite thick, but if it is too thick to blend, add a tablespoon or two of the reserved bean liquid.

ROASTED GARLIC AND GARLIC BREAD

This is a delight for serious snackers. Roasting brings out garlic's milder side. Start with a large, firm head of garlic. Place it into a small baking dish. Bake in a toaster oven or a regular oven at 375 degrees until the cloves feel soft when pressed lightly, about 25 minutes.

Then, just pick off a clove and pop it in your mouth. Or use it as a spread for bread. You can store it in the refrigerator for up to 2 weeks.

To make garlic bread, use a fork to mash peeled cloves of roasted garlic into a paste. Spread with a knife onto slices of French bread. Sprinkle with Italian seasoning if desired. Wrap tightly in foil and bake at 350 degrees for 20 minutes.

OVEN "FRIED" POTATOES

Serves 8

3 pounds medium baking potatoes
1 tablespoon vegetable oil
1 to 2 teaspoons chili powder

Preheat the oven to 425 degrees. Coat a large rimmed baking pan with vegetable cooking spray. Cut each potato in half lengthwise, then cut each half lengthwise into quarters. In a large bowl, toss together potatoes, oil, and chili powder until the potato wedges are well coated. Spread potatoes on the greased pan in one layer. Bake for about 20 minutes, or until nicely browned.

JICAMA WITH ORANGE JUICE AND MINT

Jicama, a crisp, mild-tasting root, takes on the flavor of whatever it is with. Peel it and cut it in thin slices. Arrange in a bowl and pour over fresh squeezed orange juice and fresh or dried chopped mint leaves. Toss ingredients and refrigerate for several hours.

JICAMA WITH LIME JUICE
AND MEXICAN SEASONING

Peel and slice the jicama into long narrow pieces about ¼ inch thick by ½ inch wide. Arrange on a flat pan and sprinkle the slices with fresh lime juice and Mexican seasoning or a mixture of garlic salt and chili powder. Refrigerate at least ½ hour.

CARAMEL CORN

Here is an easy Cracker Jack–like snack. Use the greater amount of popcorn if you prefer a less sweet snack.

¾ cup packed brown sugar
4 tablespoons margarine
3 tablespoons corn syrup or other liquid sweetener
¼ teaspoon salt
¼ teaspoon baking soda
¼ teaspoon vanilla extract
8 to 12 cups popped popcorn
1 cup peanuts (optional)

Combine the sugar, margarine, corn syrup, and salt in a saucepan and cook over low heat until the margarine is melted. Then continue to cook (without stirring) for 3 minutes. Add the baking soda and vanilla. Pour over popcorn and peanuts and mix until evenly coated. Bake for 15 minutes at 300 degrees. Break into pieces.

GINGERBREAD CAKE

Serves 8

This quick and delightful recipe comes from Mary Clifford, a wonderful cook. A dab of peanut butter adds a rich but mellow flavor to an old-time favorite. Let this cake cool completely before slicing or it will be crumbly.

⅓ cup margarine
⅓ cup firmly packed dark brown sugar
1½ tablespoons peanut butter
¾ cup light molasses
2 cups unsifted all-purpose flour
1 tablespoon baking powder
1 tablespoon ground ginger
1 teaspoon cinnamon
½ teaspoon ground nutmeg
Pinch of salt
¾ cup water

Preheat the oven to 325 degrees. Grease and flour an 8-inch round cake pan.

In a large bowl, cream together the margarine, brown sugar, peanut butter, and molasses. Mix together the flour, baking powder, ginger, cinnamon, nutmeg, and salt and stir into margarine mixture alternately with water until well combined. Pour batter into prepared pan. Bake about 40 minutes, or until cake tester or knife inserted in center comes out clean. Let cool completely on a wire rack.

CARROT CAKE

Serves 9

This delicious cake is also wonderfully healthy—only 9% of its calories come from fat, even with the frosting, and no cholesterol at all.

2 cups grated carrots
1½ cups raisins
2 cups water
1½ teaspoons cinnamon
1½ teaspoons allspice
½ teaspoon cloves
1 cup sugar
½ teaspoon salt
3 cups unbleached or whole wheat pastry flour
1½ teaspoons baking soda
1 cup soymilk
tofu cream frosting (see below)

Simmer the grated carrots, raisins, water, and spices in a saucepan for 10 minutes. Stir in the sugar and salt and simmer for 2 more minutes. Cool completely. Preheat the oven to 350 degrees.

In a large bowl, stir the flour and soda together. Add the cooled carrot mixture along with the soymilk and stir just to mix. Spray a 9 × 9–inch pan with nonstick spray and spread the batter in it. Bake for 1 hour. A toothpick inserted into the center should come out clean. Serve plain or frost when completely cooled.

TOFU CREAM FROSTING

Makes 1⅓ cups, enough to frost a 9 × 9–inch cake

1 cup firm tofu (½ pound)
2 tablespoons oil
2 tablespoons fresh lemon juice
3 to 4 tablespoons maple syrup
¼ teaspoon salt
½ teaspoon vanilla extract

Combine all the ingredients in a blender and blend until very smooth. Scrape the sides of the blender often with a rubber spatula to get the frosting completely smooth.

BANANA CAKE

Serves 8

2 cups unbleached all-purpose or whole wheat pastry flour
1½ teaspoons baking soda
½ teaspoon salt
1 cup raw sugar or other sweetener
⅓ cup oil
4 ripe bananas, mashed (about 2½ cups)
¼ cup water
1 teaspoon vanilla extract
1 cup chopped walnuts

Preheat the oven to 350 degrees.

Mix the flour, baking soda, and salt in a bowl. In a large bowl, beat the sugar and oil together, then add the bananas and mash them. Stir in the water and vanilla, and mix thoroughly. Add the flour mixture along with the chopped walnuts, and stir to mix.

Spread in a nonstick or lightly oil-sprayed 9-inch square baking pan, and bake for 45 to 50 minutes, or until a toothpick inserted into the center comes out clean.

PUMPKIN PIE

Serves 8

This unique recipe uses creamy, blended tofu instead of eggs and milk for a pie that is less fatty and is especially rich tasting.

1 16-ounce can pumpkin
1 teaspoon cinnamon
½ teaspoon ginger
¼ teaspoon nutmeg
⅛ teaspoon ground cloves
½ teaspoon salt
1 teaspoon vanilla extract
1 cup light brown sugar
1 tablespoon molasses
¾ pound soft tofu
1 unbaked pie shell

Preheat the oven to 350 degrees. Combine all ingredients except tofu and pie shell in a blender and blend until mixed and completely smooth. Break the tofu into chunks and add to the pumpkin mixture. Blend well. Pour the filling into the pie shell. Bake for 1 hour. Chill completely before serving.

PUMPKIN SPICE COOKIES

Makes 3 dozen cookies

These plump, moist cookies are an unusual treat. They use flaxseeds as a binder, rather than eggs. Flaxseeds are found at any health food store.

3 cups whole wheat pastry flour
4 teaspoons baking powder
1 teaspoon salt
1 teaspoon baking soda
1 teaspoon ground cinnamon
½ teaspoon grated nutmeg
1½ cups raw sugar or other sweetener
4 tablespoons flaxseeds
1½ cups water
1¾ cups solid-pack canned pumpkin
1 cup raisins

Preheat the oven to 350 degrees.

Mix the dry ingredients and set aside.

Blend flaxseeds and 1 cup water in a blender for 1 to 2 minutes, until the mixture has the consistency of beaten egg white. Add to the dry ingredients, along with the pumpkin, remaining water, and raisins. Mix until just combined.

Drop by tablespoonfuls onto a nonstick or lightly oil-sprayed baking sheet. Bake 15 minutes, or until lightly browned. Remove from baking sheet with a spatula, and place on a rack to cool. Store in an airtight container.

OLD-FASHIONED OATMEAL COOKIES

Makes 24 cookies

¼ pound margarine
½ cup firmly packed brown sugar
¼ cup sugar
1 tablespoon soy flour
½ teaspoon vanilla extract
¾ cup unbleached flour
½ teaspoon baking soda
½ teaspoon cinnamon
1½ cups rolled oats (or quick oats)
½ cup raisins

Preheat the oven to 350 degrees. Beat together the margarine and sugars until creamy. Add the soy flour and water and beat well to blend.

In a separate bowl, combine the flour, baking soda, and cinnamon. Add to the margarine mixture and mix completely. Stir in the oats and raisins.

Drop by rounded teaspoonfuls onto ungreased cookie sheets. Bake for 10 to 12 minutes, or until light golden brown.

GINGERED MELON WEDGES

Serves 6

Use cantaloupe or any other favorite melon for this recipe, which is a fast and elegant dessert.

    1 large cantaloupe
    1 scant tablespoon powdered sugar
    ½ teaspoon ground ginger
    1 tablespoon candied ginger (optional)

Cut the melon in half and seed. Then cut each half into chunks. Stir together the sugar and ground ginger. Add candied ginger if you like. Sprinkle over melon chunks and chill.

MIMOSA GRANITA

Serves 4

A refreshing, non-fat Italian ice.

1 cup juice from 3 medium oranges
½ cup sugar
1¼ cup sparkling apple juice or other sparkling fruit juice
1 tablespoon lime juice

Whisk orange juice and sugar in large bowl until sugar dissolves. Stir in sparkling juice and lime juice and pour mixture into 2 ice cube trays.

Freeze mixture until firm, at least 2 hours. If you like, you can keep frozen cubes in zipperlock plastic bags up to 1 week. Just before serving, place a single layer of frozen cubes in the bowl of a food processor fitted with a steel blade. Pulse 10 or 12 times or until no large chunks of ice remain. Scoop crystals into individual bowls. Repeat with remaining ice cubes and serve immediately.

KIWI SORBET

Serves 8

An icy treat for the hottest summer days.

4 kiwi
1 6-ounce can lemonade concentrate, thawed
2 cups water

Peel the kiwi and process in blender or food processor just until smooth. (Do not crush the seeds.) Stir in lemonade concentrate and water. Pour mixture into a metal pan. Cover with foil and freeze until firm.

Remove from freezer and let stand 10 minutes. Break into small pieces and put into food processor. Process until smooth. Pack into a plastic container and cover. Return to freezer until firm. Serve by scooping.

TROPICAL DELIGHT

Serves 3

Pureed frozen fruit makes a wonderful snack or dessert. To freeze bananas, peel and break into chunks. Freeze in a single layer on a tray, then store in an airtight container. You will find frozen pineapple and mango at the grocery store, or you can make your own by freezing canned pineapple chunks and fresh mango.

1 orange, peeled
½ cup frozen banana chunks
1 cup frozen pineapple chunks
1 cup frozen mango chunks
½ to 1 cup soymilk

Cut the orange in half and remove any seeds, then place in a blender with the remaining ingredients and process until thick and very smooth.

STRAWBERRY FREEZE

Serves 2

1 cup frozen strawberries
1 cup frozen banana chunks
½ cup unsweetened apple juice

Place all ingredients into blender and process on high speed until thick and smooth. You will have to stop the blender frequently and stir the unblended fruit to the center.

BANANA FREEZE

Serves 1

½ cup soymilk or rice milk
1 cup frozen banana chunks

Place soymilk and banana in blender and blend until thick and smooth, stopping the blender occasionally to stir the unblended fruit to the center.

For a delicious variation, add 3 pitted dates to the banana and soymilk, and process as above.

FRUIT POPSICLES

Serves 4

A fun and easy-to-make treat.

 1 ripe banana, peeled
 1 ripe peach, peeled
 6 to 8 strawberries, hulled
 1 tablespoon brown sugar
 ¼ cup soymilk
 ¼ cup fruit juice (apricot nectar, orange juice, white grape
   juice, or other juice of your choice)

Put all ingredients in a blender and process until smooth. Pour into small paper cups and place on a tray in the freezer. Add popsicle sticks when the mixture has thickened slightly. Continue to freeze until firm.

STRAWBERRY SMOOTHIE

Serves 2

Here is a thick and creamy shake that's a great, quick, and easy way to start your morning or a refreshing way to cool off on a hot summer's day. It's always a hit with kids, too.

1 frozen banana, cut into chunks (peel and wrap banana in plastic wrap before putting in freezer)
½ cup frozen strawberries or a mixture of berries and other fruit
½ cup vanilla soymilk
2 tablespoons strawberry or other fruit syrup
3 ice cubes

Place all ingredients in a blender and blend on high speed until smooth and creamy.

MINTY RED ZINGER ICED TEA

2 to 3 tea bags of Red Zinger or other robust herb tea
½ to 1 cup of fresh mint leaves (stems and all)
1 tablespoon sugar or liquid sweetener
6 cups boiling water

Place tea bags, mint leaves, and sugar in a large tea pot or pitcher, and pour in 6 cups boiling water. Let steep at least one hour, then cool in refrigerator before adding ice.

PEACH-FLAVORED
HERB ICED TEA

Follow the above directions, substituting your favorite herbed tea blend, 2 cardamom pods, a stick of cinnamon, and a sliced peach. Steep well before chilling. Strain and serve iced with a sprig of lemon balm or mint.

GREEN TEA COOLER

2-inch piece of fresh ginger root, cut into quarters
4 bags green tea
1 tablespoon sugar
4 cups of boiling water
1 bottle ginger ale

Place ginger, tea bags, and sugar in a pitcher. Pour over 4 cups of boiling water. Let steep at least 30 minutes and cool before adding ginger ale. You can make up this "base" and keep in the refrigerator. When ready to serve, add ginger ale (half tea mixture and half ginger ale is a good mixture) and serve with ice and a sprig of mint.

POWER SNACKING TIPS

- Snacking can be a healthy part of your weight control program. The healthiest snacks are very low in fat and get most of their calories from carbohydrates.
- Fruits, including dried fruits, are a terrific snack anytime.
- Sugar is less of a problem than fats in foods. But when sugar lures you to cookies or cakes, the fat they hold really can be fattening.
- The best snacks omit animal products and keep vegetable oils to a bare minimum.
- Snacks made from whole grains, such as whole-grain toast, are preferable to refined grains, which have the fiber removed.
- Your natural hunger and satiety cues will help you. Even though the snacks in this book are great for keeping the pounds off, there is no value in continuing to eat when you are no longer hungry.

- New products at health food stores make snacking easier and healthier than ever: instant soups, healthy desserts, meatless deli slices, potato chips and tortilla chips that are baked, rather than fried, etc.
- There is no need for diet sodas. The only beverage your body really needs is water, in any of the many forms it comes. Add a twist of lemon for a satisfying thirst-quencher.

HOW TO TURN YOUR FAVORITE MEALS INTO NEGATIVE CALORIE EFFECT MEALS

Don't throw away that old family recipe. Chances are you can modify it, turning a fattening food into a fat-burning food. If you cut out the fatty ingredients, you cut out a lot of useless calories. And if you boost the complex carbohydrates, you can increase your metabolism, helping you burn calories faster.

Doing that means substituting other ingredients in place of meats, dairy products, and oils, which tend to be very fatty and have no complex carbohydrate at all. Sometimes just one or two changes is all you need. Other times, you will want to give a recipe a major facelift.

In the following pages, you will find everything you need to get started: easy tips for modifying recipes, twelve sample recipes that show just how to do it, plus nutrition analyses that show how well you can get rid of extra calories and fat.

The idea is not to eliminate all the calories from your recipes. The important thing is to reduce the fat content and to build in more calorie-burning complex carbohydrate.

Reducing Fats and Oils

Fats and oils are the most calorie-dense parts of any food, with 9 calories per gram, compared to only 4 calories per gram of carbohydrate or protein.

- When you fry foods, use a nonstick pan. Avoid deep frying. If you need to add a little oil to prevent sticking, use a spray oil.
- Remember to sauté onions and garlic or other vegetables in water instead of oil. Just simmer a quarter-cup of water in a saucepan, and add the vegetables, heating for about five minutes. Or, for a little extra flavor, use vegetable stock, wine, or dry sherry.
- Steaming or baking add no fat at all. Steamer racks work wonderfully for vegetables.
- When you bake, the amount of oil or shortening in recipes is often arbitrary and can easily be cut in half or less, with no noticeable change in the

taste. Sometimes you can leave it out altogether, and substitute mashed banana, applesauce, or canned pumpkin.

- For pies, leave off the top crust. A fat-free pie crust can be made by mixing one cup of Grape-Nuts cereal and a quarter-cup of apple juice concentrate, patting the mixture into a piepan, and baking for ten minutes. Let it cool before you fill it.

- Add a sprinkle of lemon or lime to green vegetables or salads, instead of butter or oil. Delicious fat-free salad dressings are now available at all grocery stores.

- In salad dressing recipes, use water, vegetable stock, or seasoned rice vinegar instead of oil.

- In sauce recipes made with flour, liquid, and oil or fat, you can easily leave out the oil or fat. Toast the flour in a dry pan over medium heat to brown it, then add the liquid and seasonings called for in the recipe. Stir with a whisk to remove lumps, and heat, stirring constantly, until thickened.

- When nuts or peanut butter are used in recipes, they add loads of fat. You can omit them, or replace them with crunchy vegetable or fruit chunks, or Grape-Nuts cereal, depending on the type of recipe you are making.

Replacing Meat

Remember that when you replace meats, chicken, and fish with healthier ingredients, your waistline gets a real break, because all meats have substantial amounts of fat and not a speck of calorie-burning complex carbohydrate. Health food stores have products that make the switch easy.

- A full range of meatless hot dogs, burgers, and luncheon "meats" is available. Most are made from soybeans or wheat, and many are fat-free. Try different brands. Many are as tasty as they are healthful.
- To replace ground beef in spaghetti sauce, chili, sloppy joes, etc., use texturized vegetable protein (TVP). It is a fat-free soy product that is virtually indistinguishable from ground beef.
- To replace chunks of meat in stews, soups, or stir-fries, try seitan. It is an amazing product, made from wheat. Like TVP, it is fat-free and very much like the real thing.
- Tempeh is made from soybeans. It is typically marinated in soy sauce or other sauces and then grilled or barbecued. Health food stores also sell it already marinated in burger sizes. You just heat it in your toaster oven, and slip it in a bun.
- Instead of meat in tacos or chili, use pinto or black beans.

Replacing Dairy Products
Most dairy products are loaded with fat, and even the skim versions have no complex carbohydrate at all. Here are healthier choices.

- Health food stores now stock a huge range of delicious milk substitutes. Choose those with the lowest fat content. Making the transition from cow's milk to soy or rice milk is as easy as switching from whole milk to skim.

- You will also find many ice cream and yogurt substitutes, some of which are zero fat.
- To add a cheeselike taste to pizza, spaghetti, or casseroles, try nutritional yeast flakes (not baking or brewer's yeast), which are sold at all health food stores.
- For soup recipes that normally call for cream or butter, a potato provides a neat trick. Dice and boil a potato until it is just tender. Then put the potato and its cooking water into a blender, puree it, and add it to the soup.

Replacing Eggs

Eggs are loaded with fat, cholesterol, and animal protein, none of which help your waistline—or your arteries—at all.

- For binding loaves or burgers, try cooked oatmeal, mashed potato, fine bread crumbs, or tomato paste.
- For baking recipes that call for one or two eggs, just leave them out, and add a little extra water to keep the intended moisture content. If more than two eggs are called for, commercial egg replacers are available at health food stores, or substitute the following for each egg:
  - an egg-sized piece of mashed banana, applesauce, canned pumpkin, or pureed soft tofu

- a tablespoon of flaxseeds with ¼ cup water, pureed in a blender
- 2 tablespoons cornstarch
- 1 tablespoon soy flour mixed with 2 tablespoons water.

BURGERS

The Old Way

Serves 6

½ cup chopped onions
3 tablespoons olive oil
1 pound ground round
1 to 2 tablespoons barbecue sauce
¾ cup cooked bulgur
1 teaspoon salt
pepper to taste

In a large skillet, heat olive oil and sauté onions until lightly browned. In a large bowl, add the onions and remaining ingredients and mix well. Form into 6 patties and brown evenly on both sides.

A Better Way

Whole grains give this very low-fat burger a great taste and texture and pack it with "fat-fighting" carbohydrates.

Serves 6

1 cup cooked or canned butter beans
¾ cup cooked bulgur
¾ cup cooked barley
½ cup quick oatmeal, uncooked
1½ tablespoons soy sauce
1 to 2 tablespoons barbecue sauce
1 teaspoon dried basil
½ cup finely chopped onions
1 clove garlic, finely minced
1 stalk celery, chopped
1 teaspoon salt
pepper to taste

With a fork or potato masher, mash beans just slightly. They should be chunky, not pureed. Add the rest of the ingredients and form into 6 patties. Spray skillet with oil and brown patties on both sides.

NUTRITIONAL ANALYSIS PER SERVING

Old Way: *73% calories from fat; 332 calories; 14 grams protein; 7 grams carbohydrate; 27 grams fat*

Better Way: *11% calories from fat; 133 calories; 6 grams protein; 24 grams carbohydrate; 1.5 grams fat*

CHILI

The Old Way

Serves 4

1 pound ground round
2 tablespoons vegetable oil
¼ cup olive oil
1 cup chopped onions
2 minced cloves garlic
1 chopped red or green pepper
2 16-ounce cans Italian-style peeled tomatoes, chopped
½ 6-ounce can tomato paste
1 jalapeno pepper, chopped
2 tablespoons chili powder or more to taste
1 teaspoon dried oregano
2 teaspoons ground cumin
1 teaspoon cumin seeds
½ teaspoon cinnamon
¼ teaspoon allspice
⅛ teaspoon cloves
1½ cups cooked pinto or black beans
1 cup (4 oz) of grated cheddar cheese for garnish

Heat oil in large pot, add ground round and brown, drain off fat and reserve meat. Add ¼ cup olive oil to pot along with onions and garlic, saute until lightly brown. Add ground round and rest of ingredients except beans and cheese. Simmer for 1 hour, then add beans and simmer an additional 25 minutes. Garnish each serving with cheddar cheese.

A Better Way

Even "dyed-in-the-wool" chili lovers won't be able to tell that this is a *low-fat* version of a classic chili recipe.

Serves 4

1 cup dry TVP (textured vegetable protein)
1 cup boiling water
2 16-ounce cans Italian-style peeled tomatoes, chopped
½ 6-ounce can tomato paste
1 cup chopped onions
1 chopped red or green pepper
2 minced cloves garlic
1 jalapeno pepper, chopped
2 tablespoons chili powder or more to taste
1 teaspoon dried oregano
2 teaspoons ground cumin
1 teaspoon cumin seeds
½ teaspoon cinnamon
¼ teaspoon allspice
⅛ teaspoon cloves
1½ cups cooked pinto or black beans

½ cup chopped cilantro (optional garnish)
1 bunch finely chopped green onions, some green tops
   included

Pour boiling water over the TVP and let stand 5 minutes. Along with the rehydrated TVP, combine the remaining ingredients except for the beans, cilantro, and green onions. Simmer covered for about 1 hour. Add beans and cook an additional 25 minutes. Serve over brown rice, garnish with chopped cilantro and chopped green onions. Make early in the day or, better yet, the day before; the taste just gets better and better.

NUTRITIONAL ANALYSIS PER SERVING

Old Way: *60% calories from fat; 697 calories; 34 grams protein; 32 grams carbohydrate; 47 grams fat*

Better Way: *3% calories from fat; 236 calories; 19 grams protein; 31 grams carbohydrate; 0.75 gram fat*

CHOCOLATE CAKE WITH CHOCOLATE FROSTING

The Old Way

Serves 14

2 ounces chocolate
5 tablespoons boiling water
½ cup butter
1½ cups sugar
4 eggs, separated
1¾ cups cake flour
4 teaspoons baking powder
¼ teaspoon salt
½ cup whole milk
1 teaspoon vanilla

Melt chocolate, add boiling water and cool mixture. In a large bowl beat butter until soft, add sugar gradually, beat in egg yolks one at a time, then add chocolate mixture. Sift flour, baking powder, and salt into a large bowl, add to butter/chocolate mixture along with milk and vanilla.

277

Beat until smooth. Add four stiffly beaten egg whites, and fold in carefully. Bake in a greased pan 30 minutes at 350 degrees.

A Better Way

This rich-tasting cake has far less fat than the above recipe, yet it's sure to please any chocolate lover with its rich flavor and moist, light texture.

To make it almost totally fat-free follow the Fat-Free Variation and substitute a dusting of powdered sugar instead of the frosting.

Serves 10

1½ cups unbleached flour
3 tablespoons unsweetened cocoa powder
1 cup sugar
1 teaspoon baking soda
1 teaspoon salt
1 ripe banana, mashed or sliced
3 tablespoons vegetable oil
1 tablespoon vinegar
1 teaspoon vanilla
1 cup cold water

Sift the flour, cocoa, sugar, baking soda, and salt together in a large mixing bowl. In a small bowl, combine the banana, oil, vinegar, vanilla, and water and beat until blended. Add to dry ingredients, blend well and pour into an ungreased 9-inch cake pan. Bake in preheated oven for 25 minutes at 350 degrees.

FAT FREE CHOCOLATE CAKE VARIATION

This variation will give you an almost totally fat-free cake. Although it is slightly denser, it is rich tasting, wonderfully chocolatey, with just a hint of banana flavor.

Use the above recipe, but omit the oil and use a whole *large* ripe banana. Instead of the frosting dust the cake when it is cool with powdered sugar. If you place a paper lace doily over the cake and sift the powdered sugar over it, when you remove the doily you will have a pretty, decorative pattern!

RASPBERRY DREAM CAKE

For a delicious variation on chocolate cake, use the Better Way cake recipe in a 4½ by 8½–inch bread loaf pan. Cook for 35 minutes. Cool, then cut the cake in 2 or 3 layers. An easy way to do this is to encircle the cake with a two-foot piece of string or thread, making sure that it is even all around, then pull the ends of the string together until the string has "cut" through the cake. Then spread with 2 to 3 tablespoons raspberry jam and frost with Chocolate Cream Frosting.

NUTRITIONAL ANALYSIS PER SERVING

Old Way: 37% calories from fat; 313 calories; 4 grams protein; 47 grams carbohydrate; 13 grams fat

Better Way: 13% calories from fat; 254 calories; 3 grams protein; 56 grams carbohydrate; 4 grams fat

Fat-Free: 2% calories from fat; 181 calories; 2 grams protein; 41 grams carbohydrate; 0.5 gram fat

CHOCOLATE CREAM FROSTING

The Old Way

1½ cups powdered sugar
2 tablespoons unsweetened cocoa
3 tablespoons margarine, softened
2 tablespoons half-and-half
1 teaspoon vanilla

Cream together sugar, cocoa, and margarine. Add half-and-half and vanilla, beat until smooth.

A Better Way

1¼ cups powdered sugar
2 tablespoons unsweetened cocoa
2 tablespoons margarine
dash of salt
2 tablespoons plain or vanilla soymilk
1 teaspoon vanilla

Cream together the sugar, cocoa, margarine, and salt. Add the milk and vanilla to make a creamy, spreadable consistency.

LASAGNE

The Old Way

Serves 8

1 can Italian-style peeled tomatoes
2 8-ounce cans tomato sauce
1 teaspoon salt
1½ teaspoons dried oregano
1 teaspoon onion salt
1 cup minced onions
2 minced cloves garlic
⅓ cup olive oil
2 pounds ground round
1 teaspoon salt
¾ pound lasagne noodles
½ pound ricotta cheese
½ pound thinly sliced mozzarella
½ pound grated Parmesan

In a large pan combine the tomatoes, tomato sauce, salt, oregano, and onion salt. In a fry pan, sauté the onions and garlic in olive oil until golden; add ground meat and cook until meat loses it red color. Add onion meat mixture to tomato sauce and simmer gently 2½ hours.

Prepare lasagne noodles; cook 18 minutes in boiling salted water, drain and separate noodles.

In the bottom of a well-greased 9 × 13–inch baking dish, put a thin layer of sauce, then criss-cross a layer of the lasagne noodles and a layer of cheese. Repeat twice with sauce, noodles, and cheese. The final cheese layer is covered once more with sauce and a dusting of Parmesan. Bake at 350 degrees for about 40 minutes.

A Better Way
Instead of being loaded with fat this new lasagne is filled with delicious, healthy ingredients. By removing the meat and cheese, we have drastically and instantly reduced the fat and boosted the percent of carbohydrates. In place of meat we have used tasty brown rice and pinto beans to add fiber as well as increase the fat-burning carbohydrates. Aside from the healthy aspects of this revised lasagne, it takes less than half the time to prepare than the original recipe.

Serves 8

1 12-ounce package lasagne noodles
1 package frozen chopped spinach thawed and drained
1 16-ounce can of pinto beans
1 14.5-ounce can Italian-style stewed tomatoes
1 cup chopped onions
½ cup chopped red pepper

2 cups cooked brown rice
½ cup medium salsa
black pepper and garlic salt to taste
1 26-ounce can or jar of light spaghetti sauce with
    mushrooms
¼ cup minced parsley for garnish (optional)

Boil lasagne in water 18 minutes. Drain and cool. Combine spinach, beans, tomatoes, onions, red pepper, brown rice, salsa, and black pepper and garlic salt. Spread about 1 cup of the tomato sauce in the bottom of a large casserole (or two smaller 9 × 13 ones). Place half of the lasagne noodles on top of the sauce and spread with half of the bean mixture; repeat with the noodles and bean mixture. Cover with the remaining sauce. Bake covered 20 minutes at 350 degrees, remove the top and bake an additional 10 minutes or until piping hot.

NUTRITIONAL ANALYSIS PER SERVING

Old Way: *48% calories from fat; 555 cal; 26 grams protein; 12 grams carbohydrates; 30 grams fat*

Better Way: *less than 4% calories from fat; 310 calories; 14 grams protein; 64 grams carbohydrates; 1 gram fat*

CREAM OF MUSHROOM SOUP

The Old Way

Serves 4

1½ cups chicken broth
½ cup chopped onion
1 teaspoon herbs de Provence
½ pound of mushrooms, sliced
3 tablespoons butter
2 tablespoons flour
salt and pepper to taste
1 cup half-and-half
¼ cup dry sherry

In a sauce pan, combine broth, onions, herbs, and mushrooms, bring to a boil, reduce heat, cover, and simmer for 20 minutes. Melt butter in a separate pan, add flour and cook for 2 minutes; slowly add the half-and-half. Stir constantly so that no lumps form. In a blender, coarsely chop half the mushroom/broth mixture. Add both onion/mushroom mixtures to the cream mixture, stir well, and add salt and pepper. Simmer gently for 1 minute. Add sherry right before serving.

A Better Way

This full-bodied mushroom soup is not only low-fat, but is fast and easy to make as well. Pureed silken tofu gives it its creamy rich taste. Vogue Vegetable Broth is a tasty substitute for homemade vegetable stock; it comes in a powdered form and can be found in most health-food stores.

Serves 4

1½ cups vegetable broth (made from Vogue Instant
   Vegetable or Knorr Vegetarian Bouillon cubes)
½ cup chopped onion
1 teaspoon herbs de Provence
¾ pound of mushrooms, sliced
10½ ounces silken soft tofu
1 cup water (or ¾ cup for a thicker soup)
1 teaspoon low-sodium soy sauce
1 teaspoon salt
pepper to taste
¼ cup dry sherry

In a saucepan, combine broth, onions, herbs, and mushrooms and bring to a boil. Reduce heat, cover, and simmer for 20 minutes. Coarsely chop the mixture in a food processor and return it to the pan. In the food processor, blend until smooth the tofu, water, and soy sauce. Add slowly to broth/mushroom mixture, blend well. Add sherry and salt and pepper to taste.

NUTRITIONAL ANALYSIS PER SERVING

Old Way: *70% calories from fat; 209 calories; 4 grams protein; 13 grams carbohydrate; 44 grams fat*

Better Way: *23% calories from fat; 89 calories; 8 grams protein; 2 grams carbohydrate; 2 grams fat*

POTATO SALAD

The Old Way

Serves 4

2 cups sliced boiled new potatoes
½ cup French dressing
3 hard-cooked eggs, chopped
¼ cup chopped onions
½ cup chopped celery
2 tablespoons chopped olives
1 teaspoon salt
pepper to taste
½ cup mayonnaise

Boil potatoes in their jackets in a covered saucepan until they are tender. Let marinate in ½ cup French dressing. Add chopped eggs, onions, celery, and olives. Season with salt and pepper. Let stand at least an hour, then add mayonnaise. Toss carefully.

A Better Way

This potato salad is loaded with flavor and low in fat. Only 3 tablespoons of an eggless, low-fat mayonnaise (compared to a ½ cup French dressing and ½ cup mayonnaise) lowers the fat content from 41 grams to just 2 grams per serving. Carrots and peas replace the hard-cooked eggs. Whether using fresh or frozen peas, add them just before serving to keep their color bright green.

Serves 4

1 pound red potatoes, unpeeled, sliced ¼ inch thick or cut
   into pieces about the size of a walnut
2 carrots, peeled, cut lengthwise and then cut into small
   pieces
2 tablespoons dry sherry
1 tablespoon balsamic or wine vinegar
3 tablespoons eggless mayonnaise (Nayonaise)
1 teaspoon crushed dried tarragon
salt and pepper to taste
½ cup green peas
1 tablespoon roasted sunflower seeds (optional)

Cook potatoes in boiling water until just tender about 12 to 15 minutes. Remove with slotted spoon and let drain in a colander. Cook carrots in the potato water for 3 or 4 minutes, drain. Mix together sherry, vinegar, mayonnaise, herbs, and salt and pepper and pour over potatoes and carrots. Mix well and let sit at least an hour. Stir in peas and sprinkle with roasted sunflower seeds right before serving.

NUTRITIONAL ANALYSIS PER SERVING

Old Way: *76% calories from fat; 482 calories; 6 grams protein; 16 grams carbohydrate; 41 grams fat*

Better Way: *18% calories from fat; 99 calories; 3 grams protein; 21 grams carbohydrate; 2 grams fat*

STRAWBERRY MILKSHAKE

The Old Way

Serves 2

2 scoops strawberry ice cream (½ cup)
1 cup whole milk
2 tablespoons strawberry or other fruit syrup
3 medium ice cubes

Place all ingredients in a blender and blend on high speed until smooth.

A Better Way
It's hard to believe that this thick and creamy shake is low in fat. For a *totally fat-free shake*, omit the soymilk and replace with ½ cup of water instead.

Serves 2

1 frozen banana cut in chunks (peel and wrap banana in
   plastic before putting in freezer)
½ cup frozen strawberries or mixed berries
½ cup vanilla soymilk

2 tablespoons strawberry or other fruit syrup
3 medium ice cubes

Place all ingredients in a blender and blend on high speed until smooth and creamy.

NUTRITIONAL ANALYSIS PER SERVING

Old Way: *41% calories from fat; 152 calories; 5 grams protein; 17 grams carbohydrate; 7 grams fat*

Better Way: *11% calories from fat; 83 calories; 3 grams protein; 18 grams carbohydrate; 1 gram fat*

SLOPPY JOES

The Old Way

Serves 4

1 pound ground round
1 cup chopped onions
½ cup chopped celery
2 medium red or green bell peppers, coarsely chopped
¼ cup vegetable oil
1 16-ounce can tomato sauce
1 tablespoon chili powder
1 teaspoon dried mustard powder
2 tablespoons brown sugar
1 teaspoon salt
pepper to taste

Brown ground round. Remove from pan with a slotted spoon and discard fat. Add oil and sauté onions, peppers, and celery. Add the remaining ingredients including the browned ground round and mix well. Simmer 20 minutes. Serve over hamburger buns.

A Better Way

This is the same recipe minus the hamburger and oil. We have substituted TVP, textured vegetable protein (see glossary), for the ground round. For sautéing the onion, instead of using the suggested ¼ cup of oil, we use either ¼ cup wine or tomato juice. These two substitutions save a whopping 44 grams of fat, bringing the fat content from 69% to 0%!

   ¼ cup red wine, sherry, or tomato juice
   1 cup chopped onions
   ½ cup chopped celery
   2 medium red or green bell peppers, coarsely chopped
   1 cup TVP
   1½ cups boiling water
   1 16-ounce can tomato sauce
   1 tablespoon chili powder
   1 teaspoon dried mustard powder
   2 tablespoons brown sugar
   1 teaspoon salt
   pepper to taste

In a frying pan, heat the wine or tomato juice, add onions and braise, or simmer very gently covered, for 3 or 4 minutes, then add celery and peppers and simmer an additional 5 minutes. Add the TVP, boiling water, and rest of ingredients and simmer 20 minutes. Serve over hamburger buns.

NUTRITIONAL ANALYSIS PER SERVING

Old Way: *69% calories from fat; 571 calories; 21 grams protein; 15 grams carbohydrate; 44 grams fat*

Better Way: *0% calories from fat; 177 calories; 13 grams protein; 16 grams carbohydrate; 0 grams fat*

SPAGHETTI SAUCE

The Old Way

Serves 4

2 tablespoons vegetable oil
1 pound Italian sausage, casings removed (4 sausages per
   pound)
¼ cup olive oil
1 cup chopped onions
2 minced cloves garlic
2 16-ounce cans Italian-style peeled tomatoes, chopped
1 6-ounce can tomato paste
1½ teaspoon Italian herb mix
¼ teaspoon fennel seeds
salt and freshly ground black pepper to taste
⅓ cup parsley

Heat 2 tablespoons oil in large pot, add sausage, and break
up with a fork; brown, drain off fat, and reserve meat. Add
¼ cup olive oil to pot along with onions and garlic, sauté
until lightly brown. Puree one can of tomatoes, add it,
browned sausage, and rest of ingredients, except for pars-

ley, to the pot. Simmer for 30 minutes. Remove sauce from heat and stir in chopped parsley.

A Better Way

A delicious low-fat alternative to traditional spaghetti sauce, this recipe goes well with brown rice or even a baked potato as well as pasta. It's also great over broiled, thickly sliced eggplant.

Serves 4

1 tablespoon olive oil
1 pound mushrooms, chopped or sliced, tough stems
  removed
1 teaspoon fennel seeds
¼ cup red wine or sherry
1 cup chopped onions
2 minced cloves garlic
2 16-ounce cans Italian-style peeled tomatoes, chopped
1 16-ounce can tomato paste
1½ teaspoon Italian seasoning herb mix
salt and freshly ground black pepper to taste
⅓ cup parsley

Heat 2 tablespoons olive oil in large pot, add mushrooms and fennel and brown over fairly high heat. Remove mushrooms and set aside. Add ¼ cup wine or sherry to pot, heat and add onions and garlic, sauté until limp. Puree one can of tomatoes, add it as well as browned mushrooms and rest of ingredients, except for parsley, to the pot. Simmer for

30 minutes. Remove sauce from heat and stir in chopped parsley.

NUTRITIONAL ANALYSIS PER SERVING

Old Way: 59% calories from fat; 287 calories; 8 grams protein; 22 grams carbohydrate; 19 grams fat

Better Way: 21% calories from fat; 174 calories; 6 grams protein; 28 grams carbohydrate; 4 grams fat

TACOS

The Old Way

Serves 4

1 pound ground round
3 tablespoons vegetable oil
½ cup chopped onions
1 cup tomato sauce
¼ cup chopped cilantro (optional)
2 teaspoons chili powder
1 teaspoon ground cumin
1 teaspoon salt
pepper to taste
8 corn or flour tortillas
1 cup (4 oz) grated cheese
chopped tomatoes
shredded lettuce
salsa
chopped red onions

Brown ground round; remove from pan with a slotted spoon and discard fat. Add vegetable oil and sauté onions.

Add tomato sauce, cilantro (if using), chili powder, cumin, salt, and pepper to taste, mix well. Add browned ground round. Simmer 20 minutes. Serve in warmed tortillas, top with cheese, tomatoes, lettuce, salsa, and onions.

A Better Way

As with the Sloppy Joes, we have substituted TVP (textured vegetable protein), for the ground round. For sautéing the onion, instead of using the suggested ¼ cup of oil, we used either ¼ cup red wine or tomato juice. By omitting the beef, oil, and cheese, the fat content drops dramatically, but you still have a flavorful, satisfying taco!

Serves 4

⅞ cup boiling water
1 cup TVP
¼ cup red wine, sherry, or tomato juice
½ cup chopped onions
1 cup tomato sauce
¼ cup chopped cilantro (optional)
2 teaspoons chili powder
1 teaspoon ground cumin
1 teaspoon salt
pepper to taste
8 corn or flour tortillas
shredded lettuce
chopped tomatoes
salsa
chopped red onions

Pour boiling water over the TVP and set aside. In a frying pan, heat the wine or tomato juice, add onions and braise, or simmer very gently covered, for 3 minutes. Add the TVP, tomato sauce, cilantro (if using), chili powder, cumin, salt, and pepper to taste, and simmer 20 minutes. Serve in warmed tortillas, top with tomatoes, lettuce, salsa, and onions.

NUTRITIONAL ANALYSIS PER SERVING

Old Way: 63% calories from fat; 792 calories; 33 grams protein; 42 grams carbohydrate; 56 grams fat

Better Way: 17% calories from fat; 311 calories; 17 grams protein; 40 grams carbohydrate; 6 grams fat

PIZZA

The Old Way

Serves 4

¼ cup olive oil
1 cup chopped onions
2 minced cloves garlic
½ pound ground sausage meat or ground beef
1½ cups chunky tomato sauce
1 prepared pizza crust
2 cups grated cheddar cheese and 1 cup grated mozzarella,
   mixed
1 package sliced pepperoni (approximately 1 ounce per
   person)
1½ teaspoons Italian herb mix
salt and pepper to taste

Heat ¼ cup olive oil in a frying pan. Add onions and garlic, and sauté until lightly brown, then remove with a slotted spoon. In the same pan, brown sausage meat or ground beef. On a prepared pizza crust, spread the tomato sauce and add half the cheese mixture. Then add the onion mixture and meat, and the rest of the cheese. Top with slices of pepperoni and the Italian herb mix. Bake for 10–15 minutes at 425 degrees until piping hot.

A Better Way

A pizza can be delicious *and* nutritious, without the typical fattening, artery-clogging cheese, sausage, or pepperoni. The crust is almost always low in fat. A Chef Boyardee crust has less than 1 gram of fat per serving. A Pillsbury Pizza Crust has 2.5 grams, and Boboli Italian Bread Shells have 3 grams of fat. Or make your own—it's easy with this No-Fat Pizza Crust recipe (p. 304).

For the toppings, commercial tomato sauces are quick and tasty, but check the labels. Their fat content varies all the way from 0 grams of fat to 10 grams or more. The fat-free brands are delicious and much better for your waistline.

NO-FAT PIZZA CRUST

1 package active dry yeast
1 cup warm water
⅛ teaspoon sugar
3–3½ cups all-purpose flour
1 teaspoon salt
1 teaspoon dried crumbled basil (optional)
cornmeal

Combine yeast, warm water, and sugar, and stir until yeast dissolves. When small bubbles form, add to flour, salt, and optional basil. Stir or process mixture in food processor until a ball is formed, then knead for about 10 minutes. If the dough is too sticky, add more flour gradually. Lightly oil a bowl, add dough, cover and place in a warm spot. Let dough rise until double in bulk, about 30 to 50 minutes. Punch down, remove from bowl, and knead an additional minute or two. Preheat oven to 425 degrees.

On a heavy baking pan sprinkled with cornmeal, stretch the dough out to make a circle, keeping it thin in the middle and thicker around the outside edge. The heavy, black, non-stick cookie sheets make excellent pans to use. If you like crisp crusts, you may want to bake your crust for about 8 minutes before you fill it, then add toppings and finish baking an additional 7 to 10 minutes.

BUILDING YOUR PIZZA

Serves 4

This is a great way to make pizza, using the Negative Calorie Effect. Sautéing onions in wine, sherry, or tomato juice makes a fat-free topping. Some may prefer olive oil, but it adds a touch of fat.

1½ cups chunky low-fat tomato sauce
¼ cup red wine, sherry, or tomato juice (or 2 teaspoons olive oil)
1 cup chopped onions
1½ cups mushrooms, sliced about ¼ inch thick
1 medium zucchini, sliced in half lengthwise and then cut into ¼ inch thick slices
½ red pepper, cut into ¼ inch strips
1 teaspoon dried basil
1 teaspoon herbs de Provence or Italian herb mixture
½ teaspoon garlic salt
salt and freshly ground black pepper to taste

For a thicker sauce, simmer the tomato sauce on low while you are preparing the rest of the pizza.

Heat sherry, wine, or tomato juice (or olive oil) in a

frying pan. Add onions and cook until they are transparent, then add mushrooms, zucchini, red pepper, and rest of ingredients. Mix well and sauté until tender, but still crisp. Don't overcook. Spread tomato sauce on prepared pizza crust, top with sautéed vegetables, and cook about 10 minutes or until the crust is nicely browned and the topping is bubbling hot.

NUTRITIONAL ANALYSIS PER SERVING

Pizza Topping
Old Way: *77% calories from fat; 854 calories; 39 grams protein; 7 grams carbohydrate; 73 grams fat*

Better Way: *0% calories from fat; 61 calories; 2 grams protein; 7 grams carbohydrate; 0 grams fat (With olive oil: 20% calories from fat; 81 calories; 2 grams protein; 7 grams carbohydrate; 1.8 grams fat)*

FROZEN BANANA DESSERT

Serves 2

Frozen bananas make a thick and creamy dessert with no added fat. To freeze the bananas, peel and break them into pieces. Place them loosely in a covered container to freeze.

½ cup soymilk or rice milk
4–6 pitted dates
2 cups frozen banana chunks
1 teaspoon vanilla

Place the soymilk and dates into a blender and blend until the dates are in very small pieces. Add the bananas and vanilla.

Blend on high speed until thick and smooth, stopping the blender to stir any unblended fruit to the center.

Per serving: 143 calories; 3 grams protein; 30 grams carbohydrate; 1 gram fat; 46 mg of sodium; 0 mg cholesterol

ACKNOWLEDGMENTS

A special thank you to Kate Schumann and Jennifer Raymond, both of whom are experts in nutrition and recipe preparation, for providing the sections on snacks and modifying recipes.

NOTES

1 Foster, G. D., *et al.* "Controlled trial of the metabolic effects of a very-low-calorie diet: short- and long-term effects." *Am. J. Clin. Nutr.*, 1990; 51:167–72.

2 Danforth, E., Jr., Sims, E. A. H., Horton, E. S., and Goldman, R. F. "Correlation of serum tri-iodothyrone concentrations (T3) with dietary composition, gain in weight and thermogenesis in man." *Diabetes*, 1975; 24:406.

3 Spaulding, S. W., Chopra, I. J., Sherwin, R. S., and Lyall, S. S. "Effect of caloric restriction and dietary composition on serum T3 and reverse T3 in man." *J. Clin. Endocrinol. Metab.*, 1976; 42:197–200.

4 Mathieson, R. A., Walberg, J. L., Gwazauskas, F. C., Hinkle, D. E., and Gregg, J. M. "The effect of varying carbohydrate content of a very-low-caloric diet on resting metabolic rate and thyroid hormones." *Metabolism*, 1986; 35:394–8.

5 Welle, S., Lilavivathana, U., and Campbell, R. G. "Increased plasma norepinephrine concentrations and metabolic rates following glucose ingestion in man." *Metabolism*, 1980; 29:806–9.

6 Chen, J., Campbell, T. C., Junyao, L., and Peto, R. *Diet, lifestyle, and mortality in China*. Oxford University Press, Oxford, 1990.

7 de Castro, J. M., and Orozco, S. "Moderate alcohol intake and spontaneous eating patterns of humans: evidence of unregulated supplementation." *Am. J. Clin. Nutr.*, 1990; 52:246–53.

8 Kissileff, H. R., Pi-Sunyer, F. X., Segal, K., Meltzer, S., and Foelsch, P. A. "Acute effects of exercise on food intake in obese and nonobese women." *Am. J. Clin. Nutr.*, 1990; 52:240–5.

INDEX

About the Author

NEAL BARNARD, M.D., is a physician and longtime advocate for preventive medicine, higher standards in research, and improved access to medical care. In 1985 he founded the Physicians Committee for Responsible Medicine (PCRM), a nonprofit organization based in Washington, D.C. As president of PCRM, he has called for sweeping reform of federal nutrition policies. Dr. Barnard is the editor-in-chief of *Good Medicine,* PCRM's quarterly magazine, and has published editorials in the *Washington Post, Medical World News, Physicians' Weekly, USA Today, San Francisco Examiner,* and scientific publications. He is the author of several books on health and nutrition.